DATE			

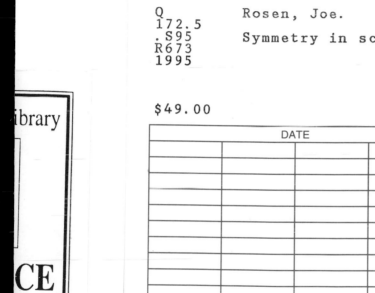

Symmetry in Science

Joe Rosen

Symmetry in Science

An Introduction to the General Theory

With 84 Illustrations

Springer-Verlag
New York Berlin Heidelberg London Paris
Tokyo Hong Kong Barcelona Budapest

Joseph Rosen
Department of Physics and Astronomy
University of Central Arkansas
Conway, AR 72035
USA

and

School of Physics and Astronomy
Tel-Aviv University
69978 Tel-Aviv
Israel

Library of Congress Cataloging-in-Publication Data
Rosen, Joseph, 1933–
  Symmetry in science: an introduction to the general theory /
Joseph Rosen.
    p. cm.
  Includes bibliographical references and index.
  ISBN 0-387-94375-7
  1. Symmetry. 2. Group theory. I. Title.
Q172.5.S95R673   1995
501′.51164—dc20   94-30622

Printed on acid-free paper.

Production coordinated by Brian Howe and managed by Natalie Johnson; manufacturing supervised by Jacqui Ashri.
Typeset by Asco Trade Typesetting Ltd., Hong Kong.
Printed and bound by Edwards Brothers, Inc., Ann Arbor, MI.
Printed in the United States of America.

9 8 7 6 5 4 3 2 1

ISBN 0-387-94375-7 Springer-Verlag New York Berlin Heidelberg

For Mira

Preface

This book is a revision and expansion of my book *A Symmetry Primer for Scientists*, published by Wiley in 1983. I expanded and revised that book mainly because, since it was published, my further quest for symmetry comprehension has led to a deeper understanding of the concepts and ideas underlying and supporting the formal theory of symmetry and its manifestation and application in science that was the main import of *A Symmetry Primer for Scientists*. Furthermore, during that period some of my ideas about symmetry in science have changed somewhat. So the expansion of the previous book was primarily by the addition of three new chapters of conceptual considerations at the end of the book. That allows the reader to choose an "application" track, whereby he or she leaves the conceptual chapters for the end, or skips them altogether, or to choose a "concept" track, in which the conceptual chapters are read before any formalism or mathematics is approached. The revision involved appending a summary of the six symmetry principles derived in the book, adding a summary section for each chapter, updating the Bibliography, and making various and numerous modifications and additions throughout.

As was its predecessor, this book is intended to fill a huge gap in the symmetry literature: Until the appearance of *A Symmetry Primer for Scientists* there was, to the best of my knowledge, not a single textbook devoted to the fundamentals of symmetry and its application in science. And since the appearance of *A Symmetry Primer for Scientists* no other such textbook has been published, again to the best of my knowledge. Indeed, that is in glaring contradiction to the undisputed importance of symmetry considerations in modern science. It is true that wherever its importance is greatly appreciated there exist textbooks on the application of symmetry in those fields, especially in quantum theory, crystallography, and chemistry. But the fact that those textbooks are almost always couched in terms of group theory and its application, rather than symmetry and its application, is symptomatic of the underlying problem this book and its predecessor are intended to alleviate: Just as technical mastery of the Schrödinger equation does not imply commensurate understanding of the physics represented by that equation, neither does proficiency in the application of group theory in some field of science imply insight into the symmetry considerations of which that group theory is the formal expression.

In fact, all too many science students, science teachers, and even scientists are uncomfortable, to say the least, with symmetry considerations. Students tend to be suspicious of symmetry arguments. They all too rarely take advantage of them and then all too often do it wrong. And when the lecturer, textbook, or thesis advisor uses a symmetry argument in a proof or to cut swiftly through to the solution of a problem, students feel cheated; something was pulled out of a hat. Science teachers often do not have a clear understanding of symmetry considerations, so they tend to avoid them, except when a symmetry argument is very obvious and compelling. And even those teachers who do understand and appreciate symmetry considerations rarely convey to their students a feeling for, if not some understanding of, the principles of symmetry and their generality and importance. Then we have the scientists, too many of whom tend to be doubtful of symmetry considerations, especially their own, because they are uncertain about how to use them correctly. And those doubtful scientists might at the same time be very proficient at the group-theoretical techniques involved in the application of symmetry in their own fields.

So what is sorely needed is a basic textbook on symmetry and its application in science, starting from scratch and presenting the material in an orderly manner, with problems for the reader to solve and a bibliography. Such a book should be able to serve as the primary text for a course on symmetry, as a supplementary text for science courses in which symmetry considerations play an important role, and as a self-study text for scientists, science teachers, and advanced science students who want to fill in what they may have missed and increase their symmetry sophistication. Detailed specific applications need not be included, since they can be found in abundance elsewhere. As you might suspect, this book attempts to fill just that role.

As for starting from scratch, the conceptual presentation of symmetry, the formal presentation of symmetry, and the presentation of group theory (the mathematical language of symmetry) and of the principles of symmetry do indeed do so. However, for some of the additional material and for examples and problems throughout the book the reader is assumed to possess a certain background familiarity with various mathematical subjects and with various physical phenomena. For example, some familiarity with elementary algebra, complex numbers, geometric concepts, linear algebra, or ordinary differential equations is assumed at various points. And, especially for the material on quantum systems, a more than introductory understanding of quantum theory is required.

Thus, in keeping with its many roles, not all of the book is accessible to all readers. The undergraduate science student should be able to handle all the core material (see below) except the quantum material and some of the examples and problems. The advanced graduate students should find almost all the material, examples, and problems accessible, as should the science teacher and scientist. And any reader, upon returning to this book after gaining experience and insight in science, should find that more of it has become accessible and some of it has become clearer.

The book is structured as follows. Chapter 1 serves as a brief gateway to symmetry by introducing the notion of symmetry in its generality. Chapters 2 and 3 are an introduction to group theory, the mathematical language of symmetry. They present a reasonable dose of group theory and certain other mathematical ideas and supply the reader with the mathematical ideas and language necessary for the succeeding chapters, where a symmetry formalism is developed in group-theoretical language, the language most suitable for it. For Chapters 2 and 3 to serve as a useful introduction to group theory they go somewhat further than is strictly necessary for the purpose of this book.

Chapter 4 starts the development of a general symmetry formalism, which is continued in the following chapters. It presents a formalism applicable to all, not necessarily even physical, systems. A special section on symmetry in quantum systems is included. Chapter 5 teaches the theory of application of symmetry in science. It contains some discussion of a somewhat philosophical nature, but most of it is utilitarian, and numerous examples are included. Those chapters comprise the core of the "practical" part of the book. The remainder of the "practical" part consists of Chapters 6 and 7, which require a more sophisticated reader than does most of the core material, Chapter 7 more than Chapter 6. Those chapters can be considered somewhat "extracurricular." They are not necessarily less important, however, and the first four sections of Chapter 7, discussing symmetry of the laws of nature and of initial and final states, symmetry in processes, and conservation, definitely should be read by the sufficiently sophisticated reader.

The "conceptual" part of the book consists of Chapters 8–10, which are a conceptual discussion of symmetry and its manifestation and application in science. Chapter 8 discusses symmetry at its most general, while symmetry in science is considered in Chapters 9 and 10.

Numerous problems are included in this book, following almost every section of all chapters except 1 and 6. The reader should, of course, attempt to solve them in order to enhance the learning process and for self-testing. The problems of Chapters 2 and 3, a few of which extend the presentation, and the problems of Chapters 8–10 should all be accessible to whomever the chapters are accessible to. The problems of Chapters 4, 5, and 7, on the other hand, are not all accessible to every reader. Some of the problems purposely leave leeway for the reader's interpretation and thus do not have unique solutions.

A rather extensive bibliography is included, whose principal purpose is to offer sources for parallel, supplementary, complementary, and subsequent reading. Being very self-contained, this book rarely makes direct reference to the Bibliography. When it does, a reference is designated by square brackets, [...].

For the person desiring preparatory reading to this book the cupboard is almost bare. As a non-textbook introduction to symmetry I dare suggest my own *Symmetry Discovered: Concepts and Applications in Nature and Science* [S22]. It is the only one I know of at its level and, like the present book, was also intended to fill a gap in the symmetry literature. For a more advanced non-textbook introduction to symmetry I highly recommend Hermann Weyl's modern classic, *Symmetry* [S31].

A symmetry course based on this book might be structured as follows. If you, the instructor, prefer the "application" track, with the conceptual material relegated to the end or left out altogether, then start with Chapter 1 (What Is Symmetry?). Follow with Chapters 2 (The Mathematics of Symmetry: Group Theory) and 3 (Group Theory Continued), if the students need an introduction to group theory. For a thorough introduction use all of both chapters. For a brief introduction that is sufficient for the rest of the course the essential sections, are in Chapter 2: 2.1. The Group Concept; 2.2. Mapping; 2.3. Isomorphism; 2.4. Equivalence Relation; 2.6. Subgroup; and in Chapter 3: 3.5. Generators; 3.7. Permutations, Symmetric Groups; and the material on equivalence class in Section 3.1. If the students have already been sufficiently introduced to group theory, mapping, and equivalence relation, Chapters 2 and 3 may be skipped, unless a rapid review is desired. In no case should one attempt to teach the subsequent chapters to students who are not sufficiently familiar with group theory, mapping, and equivalence relation.

Then present Chapters 4 (Symmetry: The Formalism) and 5 (Application of Symmetry). The last section of each chapter is only for students with a good formal background in quantum theory. After that Chapters 6 (Approximate Symmetry and Spontaneous Symmetry Breaking) and 7 (Symmetry in Processes, Conservation, and Cosmic Considerations) are for your picking and choosing. For these chapters a rather advanced science background is expected of the students.

At this point in the "application" track, if time and interest allow, the conceptual foundations of symmetry may be considered by presenting Chapters 8 (Symmetry: The Concept), 9 (Symmetry in Science), and 10 (More Symmetry in Science). This material can and should be pared down according to your taste and interest.

If you, the instructor, choose the "concept" track rather than the "application" track, preferring to start with the conceptual foundations of symmetry and its application in science and only then develop the formalism for application, start with Chapter 1 (What is Symmetry?) and continue with Chapters 8 (Symmetry: The Concept), 9 (Symmetry in Science), and 10 (More Symmetry in Science), adapting this material to your taste, interest, and needs as you find appropriate. Then present Chapters 2–7 as described above for the "application" track.

The most important feature of this book, it seems to me, is the statement and rigorous derivation of six principles of symmetry: the equivalence principle (Section 5.2); the symmetry principle (Section 5.3); the equivalence principle for processes (Section 7.2); the symmetry principle for processes (Section 7.2); the general symmetry evolution principle (Section 7.2); and the special symmetry evolution principle (Section 7.3). I know that the equivalence principle, the symmetry principle, and the special symmetry evolution principle have been stated previously in one version or another [A17, A5, M48]. About the other three principles I do not know. In any case, I am sure this (and in *A Symmetry Primer for Scientists*, the present book's predecessor) is the first time all principles have been presented together, in their full generality, and within a coherent framework. And I am sure also that this is the first time the principles have been rigorously derived from a premise as fundamental as the very existence of science. The six principles are summarized in the Summary of Principles, following Chapter 10.

I would like to express my thanks to the Department of Physics of The Catholic University of America, and especially to Larry Fagg, Jim Brennan, and Jack Leibowitz, for a most fruitful visit during 1990–1993, when this book was written.

Bethesda   Joe Rosen

Contents

CHAPTER ONE

What Is Symmetry?

Everyone has some idea of what *symmetry* is. We recognize the bilateral symmetry of the human body, of the bodies of many other animals, and of numerous objects in our environment. We consider a scalene triangle to be completely lacking in symmetry, while we see symmetry in an isosceles triangle and even more symmetry in an equilateral triangle. That is only for starters. Any reader of this book can easily point out many more kinds and examples of symmetry. For an elementary introduction to symmetry see my book [S22], and for a more advanced introduction, see Weyl [S31]. Escher has given us fantastic illustrations of geometric (and color) symmetry, to be enjoyed in [M16, M17, M40, and M55]. And see also [S13, S30, M25, and M35].

In science, of course, our recognition and utilization of symmetry is often more sophisticated, sometimes very much more. But what symmetry actually boils down to in the final analysis is that the situation possesses the possibility of a change that leaves some aspect of the situation unchanged.

A bilaterally symmetric body can be reflected through its midplane, through the (imaginary) plane separating the body's two similar halves. Think of a two-sided mirror positioned in that plane. Such a reflection is a change. Yet the reflected body looks the same as the original one; it coincides with the original: the reflected right- and left-hands, paws, or hooves coincide, respectively, with the original left and right ones, and similarly with the feet, ears, and so on. See Fig. 1.1.

For the triangles let us, for simplicity, confine ourselves to rotations and reflections within the plane of the figures. Then a rotation is made about a point in the plane, while a reflection is made through a line in the plane. An infinite number of such changes can be performed on any triangle. But for an equilateral triangle there are only a finite number of them that can be made on it and that nevertheless leave its appearance unchanged, i.e., rotations and reflections for which the changed triangle coincides with the original. They are the rotations about the triangle's center by 120° and by 240° and the reflections through each of the triangle's three heights, five changes alto-

1

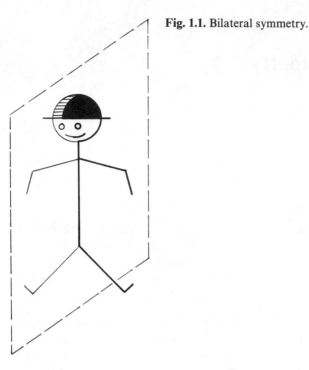

Fig. 1.1. Bilateral symmetry.

gether. (For the present we do not count rotations by multiples of 360°, which are considered to be no change at all.) See Fig. 1.2.

Although an infinity of planar rotations and reflections can be performed also on any isosceles triangle, there is only a single such change that preserves the appearance of an isosceles triangle, that leaves the triangle coinciding with itself. It is the reflection through the height on its base. See Fig. 1.3. And a scalene triangle cannot be made to coincide with itself by any planar rotation or reflection (not counting rotations by multiples of 360°). See Fig. 1.4.

We stated above that symmetry is in essence that the situation possesses the possibility of a change that nevertheless leaves some aspect of the situation unchanged. That can be concisely formulated as the following definition of symmetry:

Symmetry is immunity to a possible change.

When we do have a situation for which it is possible to make a change under which some aspect of the situation remains unchanged, i.e., is immune to the change, then the situation can be said to be *symmetric under the change with respect to that aspect*. For example, a bilaterally symmetric body is symmetric under reflection through its midplane with respect to appearance. Its external appearance is immune to midplane reflection. (The arrangement of its internal organs, however, most usually does not have this symmetry. The human heart, for instance, is normally left of center.) For very simple

Fig. 1.2. Changes bringing equilateral triangle into coincidence with itself.

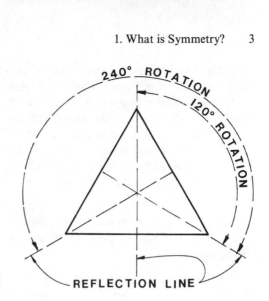

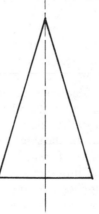

Fig. 1.3. Change bringing isosceles triangle into coincidence with itself.

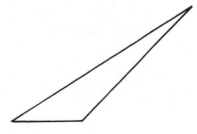

Fig. 1.4. No change brings scalene triangle into coincidence with itself.

animals their bilateral symmetry might also be with respect to physiological function. That is not true for more complex animals.

An equilateral triangle is symmetric with respect to appearance under the rotations and reflections we mentioned above. An isosceles triangle is symmetric with respect to appearance under reflection through the height on its base. But a scalene triangle is not symmetric with respect to appearance under any planar rotation or reflection.

Note the two essential components of symmetry:

1. *Possibility of a change.* It must be possible to perform a change, although the change does not actually have to be performed.

2. *Immunity.* Some aspect of the situation would remain unchanged, if the change were performed.

If a change is possible but some aspect of the situation is not immune to it, we have *asymmetry*. Then the situation can be said to be *asymmetric under the change with respect to that aspect.* For example, a scalene triangle is asymmetric with respect to appearance under all planar rotations and reflections. All triangles are asymmetric under 45° rotations with respect to appearance. An equilateral triangle is symmetric with respect to appearance under 120° rotation about its center, while isosceles triangles do not possess that symmetry; they are asymmetric under 120° rotations with respect to appearance. While a triangle might be symmetric or asymmetric with respect to appearance under a given rotation or reflection, all triangles are symmetric under all rotations and reflections with respect to their area; rotations and reflections do not change area. On the other hand, all plane figures are asymmetric with respect to area under dilation, which is enlargement (or reduction) of all linear dimensions by the same factor. The area then increases (or diminishes) by the square of that factor.

After that brief introduction to symmetry you should at this point choose between the "application" track and the "concept" track. If you prefer to postpone the conceptual discussion, or to skip it altogether, you should simply continue to the next chapter. The conceptual material is presented in the last three chapters, Chapters 8–10. On the other hand, if you choose to take the path through the conceptual thicket, then you should jump to Chapters 8–10 at this point and return to Chapter 2 afterward.

CHAPTER TWO

The Mathematics of Symmetry: Group Theory

The conceptual approach to symmetry, which is the subject of Chapters 8–10, is the best approach for understanding the concepts involved in symmetry and the significance of symmetry in science. (The reader choosing the "concept" track will have finished reading those chapters at this point.) Such understanding is, indeed, very important.

But when it comes to *applying* symmetry considerations in science, the conceptual approach can only go so far, which is not very far. And especially when the applications are quantitative, the conceptual approach is simply incapable of supplying the necessary tools. So for the application of symmetry in science it is necessary to develop a general formalism of symmetry. That formalism is developed in Chapters 4–7. The formalism is couched in the mathematical language of symmetry, which is group theory. Now, that might require the student of symmetry to learn some new mathematics. But it is unavoidable. Just as geometry is the appropriate mathematics for investigating space and calculus the appropriate mathematics for investigating motion, so is group theory the mathematics of symmetry. And just as one cannot progress in the investigation of space and motion without geometry and calculus, so is group theory essential for the further study of symmetry.

Thus in this chapter and in the next, group theory and other mathematical concepts that are essential for further serious study of symmetry are discussed. If you feel you are already familiar with the material, you might use these chapters for a review. Although it is desirable to master both chapters before continuing on to the next chapters, that is not absolutely essential. If you want to abbreviate your introduction to group theory, the sections that must not be missed are, in Chapter 2: 2.1. The Group Concept; 2.2. Mapping; 2.3. Isomorphism; 2.4. Equivalence Relation; 2.6. Subgroup; and in Chapter 3: 3.5. Generators; 3.7. Permutations, Symmetric Groups; and the material on equivalence class in Section 3.1. But if you want to understand group theory, you should definitely study all of both chapters.

This and the next chapter are only an introduction to group theory. There is very much more, and there are many books to help the interested student in his or her quest for further understanding. The Bibliography at the end of this book lists a considerable number of books for more advanced study. As you might suspect, a more advanced study of symmetry than is presented in this book requires a more advanced study of group theory than these two chapters offer. The Bibliography lists many books that take up the study of symmetry more or less where this book leaves off, and many of them contain correspondingly more advanced presentations of group theory.

If we were studying linear vector spaces, we could start by reviewing the familiar properties of ordinary vectors (2-vectors and 3-vectors), and, except for possible complications due to complex numbers and infinite dimensionality, we could use ordinary vectors as an excellent reference model throughout our presentation to help us keep our feet on the ground. But what can we do if we are studying group theory? How can we keep our feet on the ground? My own experience has led me to believe that the best solution is to leave the ground and start flying. So we will start with a completely abstract approach to groups and, with the help of examples and problems, try to develop a firm understanding of what is going on. (After all, ordinary vectors are also abstractions. It is only a matter of familiarity.)

This and the next chapter are self-contained and start very much from scratch. For the theory it would be useful to know a little set theory, but that is not essential. For the examples and problems some elementary mathematical knowledge is assumed, including concepts from algebra, functions, geometry, analytic geometry, complex numbers, and linear algebra.

2.1. The Group Concept

A *group* is, first of all, a *set* of *elements*. In a rigorous presentation one first studies some set theory before approaching group theory. (You might find it interesting to read some of the chapter on set theory in a book on abstract algebra.) However, for our purposes it is sufficient to know that a set is a collection and the elements of a set are whatever comprise the collection. An element of a set is said to be a member of the set, or to belong to the set. Besides whatever property the elements of a set have that makes them belong to the set, they need have no additional properties. For a set to be a group, however, very definite additional properties are needed, as we will see in the following discussion.

The number of elements of a set or group is called its *order*. It may be finite, denumerably infinite, or nondenumerably infinite. You know what finite means, and you most likely have a reasonable idea of what infinite means. *Denumerably* infinite means that the elements can be labeled by the natural numbers, 1, 2, 3, ... So the set of natural numbers is by definition denumerably infinite. The set of all integers (positive, negative, and zero) is also denumerably infinite, and so is (believe it or not) the set of all rational numbers (numbers expressible as the ratio of two integers). *Nondenumerably* infinite means that the elements cannot be labeled by the natural numbers. The set of all real numbers, for example, is of nondenumerably infinite order. (You might want to read about infinities, comparing infinities, greater and lesser infinities.)

We use capital letters to denote sets and groups and small letters for elements, in general. Curly brackets, $\{\ \}$, denote the set of all elements indicated or defined within them. Thus the equation

$$S = \{a, b, c, d\}$$

means that the set denoted by S consists of the elements denoted by a, b, c, d. The sign $=$ in such equations means that the set on the left-hand side and that on the right are one and the same. Another example is

$$P = \{\text{all real numbers } a \text{ such that } -1 < a < +1\},$$

which means that the set P consists of all real numbers greater than -1 but less than $+1$.

The symbol $\in$ is used to indicate membership in a set or group. The statement

$$w \in U$$

means that the element w is a member of the set U.

Another use of the sign $=$ is to relate elements of a set, as in the equation

$$a = b.$$

Here the $=$ sign means that the two elements are one and the same element.

For a set to be a group it must be endowed with a *law of composition* (also called *law of "multiplication"*; note the quotes; this is not ordinary multiplication, so "composition" is preferable to "multiplication" to avoid confusion). This means that any two elements of the set can be combined (without specifying just how that is to be done). In fact, any pair of elements can be combined in two ways. If a, b are elements of G, the two compositions (or "products") of a and b according to the law of composition of set G are denoted ab and ba. (It is fortunate in terms of notation that composition involves only two ways of combining!)

But in itself composition does not yet a group make. To be a group, a set G together with its law of composition must satisfy the following four conditions:

1. *Closure.* For all a, b such that $a, b \in G$, we have

$$ab, ba \in G.$$

This means that for all pairs of elements of G both compositions are themselves elements of G. Another way of stating this is that the set G is closed under composition.

2. *Associativity.* For all a, b, c such that $a, b, c \in G$, we have

$$a(bc) = (ab)c.$$

This means that in the composition of any three elements the order of combining pairs is immaterial. Thus one can evaluate abc by first making the composition $bc = d$ and then forming ad, corresponding to $a(bc)$, or one can start with $ab = f$ and then make the composition fc, corresponding to $(ab)c$. The results must be the same element of G, which can thus be unambiguously denoted abc. In short, one says that the composition is associative. It then follows that associativity holds for composition of any number of elements.

    3. *Existence of Identity.* G contains an element e, called an *identity* element, such that

$$ae = ea = a$$

for every element a of G. The characteristic property of an identity element is, therefore, that its composition in either way with any element of G is just that element itself.

    4. *Existence of Inverses.* For every element a of G there is an element of G, denoted a^{-1} and called an *inverse* of a, such that

$$aa^{-1} = a^{-1}a = e.$$

In other words, for every element of G there is an element whose composition with it in either way is an identity element. (It might happen that $a^{-1} = a$ for some or all elements of a group.)

    That ends the definition of a group. To summarize, a group is a set endowed with a law of composition such that the conditions of: (1) closure; (2) associativity; (3) existence of identity; and (4) existence of inverses hold. A group is *abstract* if its elements are abstract, i.e., if we do not define them in any concrete way.

    In general $ab \neq ba$ in a group. (Otherwise, why insist on the two ways of composition?) But it might happen that certain pairs of elements a, b of G do obey $ab = ba$. Such a pair of elements is said to *commute*. From the definition of a group it is seen that an identity element e commutes with all elements of a group and that every element commutes with its inverse. Obviously, every element of a group commutes with itself. If all the elements of a group commute with each other, i.e., if $ab = ba$ for all elements a, b of G, the group G is called *commutative* or *Abelian* (after the Norwegian mathematician Niels Henrik Abel, 1802–1829—but don't conclude that everyone who studies group theory dies so young).

    In condition 3 we demand the existence of an identity element but do not demand that it be unique. And in condition 4 and the preceding paragraph we carefully refrain from referring to *the* identity in order not to imply uniqueness. However, all that deviousness was only to allow us the pleasure of *proving* that the identity is unique, as it in fact is. This will be our first example of a group-theoretical proof. Please note, and follow our example in your own group-theoretical proofs, that we will be very careful to justify each operation and each equation by reference to the definition of a group. That is especially important, since we are using familiar notation (multiplication, parentheses, equality) but assigning it new significance, and one might have a tendency to revert to old ways.

To prove uniqueness of the identity, we assume the opposite, that more than one identity element exist, and show that this leads to a contradiction. Denote any two of the assumed identity elements by e' and e''. By condition 3 we then have

$$e'a = ae' = e''a = ae'' = a$$

for every element a of G. Evaluate the relation $e'a = a$ for the specific element $a = e''$. We obtain $e'e'' = e''$. Now evaluate $ae'' = a$ for $a = e'$. We get $e'e'' = e'$. Comparing these two results, we have $e' = e''$, which is in contradiction to our assumption that e' and e'' are different. That proves the uniqueness of the identity.

Although we do not require it in condition 4, the inverse of an element is unique; i.e., for every element a of G there is only one element, denoted a^{-1}, such that

$$aa^{-1} = a^{-1}a = e.$$

As an additional example of group-theoretical proof we prove that statement. We again assume the opposite and show that it leads to a contradiction. So assume that element a has more than one inverse and denote any two of them by b, c. Then by condition 4

$$ab = ba = ac = ca = e.$$

Now according to the associativity condition

$$c(ab) = (ca)b.$$

By the equation before last the left-hand side of the last equation is

$$c(ab) = ce = c,$$

while the right-hand side is

$$(ca)b = eb = b.$$

(We use the property of the identity here.) Therefore $b = c$, and we have a contradiction to our assumption that b and c are different.

From the definition of the inverse and its uniqueness it is clear that the inverse of the inverse is the original element itself, i.e., the inverse of a^{-1} is a. That is written

$$(a^{-1})^{-1} = a.$$

The inverse of a composition of elements is the composition in opposite order of the inverses of the individual elements. In symbols this means that

$$(ab)^{-1} = b^{-1}a^{-1},$$

$$(abc)^{-1} = c^{-1}b^{-1}a^{-1},$$

etc.

This is verified directly. For example, we show that $b^{-1}a^{-1}$ is indeed the inverse of ab by forming their composition and proving it is the identity

$$(ab)(b^{-1}a^{-1}) = a(bb^{-1})a^{-1} \quad \text{(by associativity)}$$
$$= aea^{-1} \quad \text{(by inverse)}$$
$$= (ae)a^{-1} \quad \text{(by associativity)}$$
$$= aa^{-1} \quad \text{(by identity)}$$
$$= e \quad \text{(by inverse).}$$

The verification of

$$(b^{-1}a^{-1})(ab) = e$$

is done similarly.

We define powers of elements by

$$a^2 = aa,$$
$$a^3 = a^2a = aa^2 = aaa,$$
$$\text{etc.,}$$

and

$$a^{-2} = a^{-1}a^{-1} = (a^2)^{-1},$$
$$\text{etc.}$$

Then the usual rules for exponents are largely applicable, except that non-commutativity must be kept in mind. For example,

$$(ab)^2 = (ab)(ab) = abab \neq a^2b^2.$$

The *structure* of a group is the statement of the results of all possible compositions of pairs of elements. For finite-order groups that is most clearly done by setting up a *group table*, similar to an ordinary multiplication table (Fig. 2.1). To find ab we look up a in the left column and b in the top row; the composition ab is then found at the intersection of the row starting with a and the column headed by b. For the composition ba it is b that is on the left and a at the top. When the symbols for the group elements have the same

e	$\cdots$	a	$\cdots$	b	$\cdots$
$\vdots$	$\ddots$	$\vdots$		$\vdots$	
a	$\cdots$	a^2	$\cdots$	ab	$\cdots$
$\vdots$		$\vdots$	$\ddots$	$\vdots$	
b	$\cdots$	ba	$\cdots$	b^2	$\cdots$
$\vdots$		$\vdots$		$\vdots$	$\ddots$

Fig. 2.1. Group table.

ordering in the top row (left to right) as in the left column (up to down), a group table will be symmetric under reflection through the diagonal if and only if the group is Abelian. That is easily seen from the figure. Note that by reordering the rows or columns or by changing the symbols denoting the group elements a group table can be made to look different while still describing the same group. That must be taken into account when comparing group tables. Thus two group tables describe the same group, i.e., express the same structure, if they can be made identical by reordering rows and columns and redenoting elements.

For groups of infinite order a group table is obviously impossible. Instead, the results of all possible compositions must be expressed by a general rule.

A concrete example of an abstract group, i.e., a group of concrete elements with a concretely defined law of composition, having the same structure as an abstract group, is called a *realization* of that abstract group. Such realizations might be, for example, groups of numbers, matrices, rotations, or other geometric transformations.

We now consider all abstract groups of orders 1 to 5 and one abstract group of order 6 and present examples, or realizations, for each but the last.

ORDER 1. There is only one abstract group of order 1, denoted C_1, and it is the trivial group consisting only of the identity element e. It is Abelian. One realization of the group is the number 1 and ordinary multiplication as the composition. Another realization is the number 0 and the composition of ordinary addition.

ORDER 2. There is only one abstract group of order 2, denoted C_2. It is Abelian. It consists of the identity element e and one other element a, which must then be its own inverse, $aa = e$. The group table is shown in Fig. 2.2. A realization of the group is the set of numbers $\{1, -1\}$ (the two square roots of 1) under ordinary multiplication. The number 1 serves as the identity, while the number -1 is its own inverse, $(-1) \times (-1) = 1$. Another, geometric, realization is the set consisting of the transformation of not doing anything, called the *identity transformation*, and the transformation of mirror reflection (with a two-sided mirror), the composition being consecutive reflection. (Transformations are discussed more thoroughly in Chapter 4.) The identity transformation serves as the group identity element, and the reflection transformation is its own inverse, since two consecutive reflections in the same mirror bring the situation back to the way it was originally. Another realization is the set of rotations about a common axis by $0°$, which is

$$
\begin{array}{c|c}
e & a \\
\hline
a & e
\end{array}
$$

Fig. 2.2. Group table of C_2.

$$
\begin{array}{c|cc}
e & a & b \\
\hline
a & b & e \\
b & e & a
\end{array}
$$

Fig. 2.3. Group table of C_3.

another way of expressing the identity transformation, and by 180°, with the composition of consecutive rotation. Note that the group elements here are not the orientations of 0° and 180°, but the *rotations* through those angles. Rotation by 0° is the identity element. Rotation by 180° is its own inverse, since two consecutive rotations by 180° about the same axis give a total rotation by 360°—no rotation at all—the identity element.

ORDER 3. There is only one abstract group of this order, denoted C_3, and it is Abelian. It consists of the identity e and two more elements a and b, each of which is the inverse of the other,

$$ab = ba = e,$$

and each of which composed with itself gives the other,

$$aa = b, \qquad bb = a.$$

Figure 2.3 is the group table. One realization is the set of complex numbers $\{1, e^{2\pi i/3}, e^{4\pi i/3}\}$ (the three third roots of 1) under multiplication. The number 1 serves as the identity element, and the other two numbers can respectively correspond to either a, b or b, a. Another realization is the set of rotations about a common axis by $\{0°$ (the identity transformation), $120° (= 360°/3)$, $240° (= 2 \times 120°)\}$, with composition of consecutive rotation. The identity transformation of rotation by 0° corresponds to the identity element, while the other two rotations can respectively correspond to either a, b or b, a.

ORDER 4. There are two different abstract groups of order 4. Both are Abelian. One of them has a structure similar to that of the abstract group of order 3 and is denoted C_4. Its group table is shown in Fig. 2.4. It can be realized by the set of numbers $\{1, i, -1, -i\}$ (the four fourth roots of 1) under multiplication. Another realization is the set of rotations about a common axis by $\{0°$ (the identity transformation), $90° (= 360°/4)$, $180° (= 2 \times 90°)$, $270° (= 3 \times 90°)\}$, with consecutive rotation as composition.

$$
\begin{array}{c|ccc}
e & a & b & c \\
\hline
a & b & c & e \\
b & c & e & a \\
c & e & a & b
\end{array}
$$

Fig. 2.4. Group table of C_4.

e	a	b	c
a	e	c	b
b	c	e	a
c	b	a	e

Fig. 2.5. Group table of D_2.

The other group of order 4, denoted D_2, has the group table of Fig. 2.5. In this group each element is its own inverse,

$$aa = bb = cc = e.$$

To obtain a realization we imagine two perpendicular, intersecting two-sided mirrors A and B and their line of intersection C as in Fig. 2.6. The results of the transformations of reflection through each mirror and rotation by 180° about their line of intersection are shown in cross section in Fig. 2.7. The set of transformations consisting of those transformations and the identity transformation, with consecutive transformation as composition, forms a group and is a realization of D_2.

ORDER 5. There is only one abstract group of order 5, denoted C_5, and it is Abelian. Its structure is similar to that of C_4. Figure 2.8 is its group table. One realization is the set of five fifth roots of 1 under multiplication. Another realization is the set of rotations $\{0°, 72° (= 360°/5), 144° (= 2 \times 72°), 216° (= 3 \times 72°), 288° (= 4 \times 72°)\}$, with consecutive rotation as composition.

All groups of orders 1 to 5 are Abelian. One might be tempted to guess by induction that all finite-order groups are Abelian. (This is the kind of "induction" by which one "proves" that 60 is divisible by all natural numbers: it is divisible by 2, by 3, by 4, by 5, by 6, and so on.) That is false. It turns out that the lowest-order non-Abelian group is one of the two order-6 groups and is

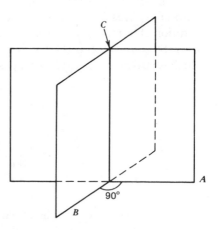

Fig. 2.6. Two perpendicular intersecting mirrors A and B and their line of intersection C.

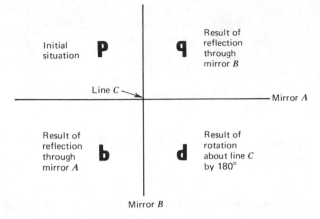

Fig. 2.7. Reflections in perpendicular mirrors A and B and rotation by 180° about their line of intersection C, shown in cross section.

e	a	b	c	d
a	b	c	d	e
b	c	d	e	a
c	d	e	a	b
d	e	a	b	c

Fig. 2.8. Group table of C_5.

denoted D_3. To dispel any lingering impression that all finite-order groups are Abelian, we display its group table in Fig. 2.9.

Now for some infinite-order groups. We present examples in terms of realizations, because the general rule for composition is most easily expressed in the context of a realization.

1. The set of all integers (positive, negative, and zero) under addition forms a group. (1) Closure holds, since the sum of any two integers is an integer. (2) Addition of numbers is associative. (3) The identity of addition is the number 0, which is an integer. (4) The inverse of any integer a is its negative $-a$, which is also an integer. The group is Abelian, since addition is commutative.

e	a	b	c	d	f
a	b	e	f	c	d
b	e	a	d	f	c
c	d	f	e	a	b
d	f	c	b	e	a
f	c	d	a	b	e

Fig. 2.9. Group table of D_3.

This set of numbers, however, does not form a group under ordinary multiplication. (1) Closure holds, since the product of any two integers is an integer. (2) Multiplication of numbers is associative. (3) The identity of multiplication is the number 1, which is an integer. But it is condition 4 that is not satisfied, since the multiplicative inverse of any integer a, which is its reciprocal $1/a$, is not in general an integer; for $a = 0$ it is not even defined.

2. The set of all nonzero rational numbers does form a group under multiplication. (1) The product of any two rational numbers is rational. (2) Multiplication is associative and even commutative, so the group is Abelian. (3) The identity of multiplication, the number 1, is rational. (4) The multiplicative inverse of any rational number a ($\neq 0$) is its reciprocal $1/a$, which is also rational, since if a is the ratio of two integers, so is $1/a$. (This set together with the number 0 forms a group under addition.)

3. The set of all $n \times n$ matrices forms a group under matrix addition. (1) The sum of any two $n \times n$ matrices is an $n \times n$ matrix. (2) Matrix addition is associative. It is also commutative, so the group is Abelian. (3) The identity of addition is the zero $n \times n$ matrix, which is a member of the set. (4) The inverse of any $n \times n$ matrix is its negative, which is also an $n \times n$ matrix. The same set of matrices fails to form a group under matrix multiplication. It has trouble with condition 4.

4. However, the set of all *nonsingular* $n \times n$ matrices does form a group under matrix multiplication. (1) The product of any two nonsingular $n \times n$ matrices is a nonsingular $n \times n$ matrix. (2) Matrix multiplication is associative. It is not commutative, so the group is non-Abelian (except for $n = 1$). (3) The identity of matrix multiplication is the unit $n \times n$ matrix, which is nonsingular. (4) The inverse of any nonsingular $n \times n$ matrix is its matrix inverse, which exists and is a nonsingular $n \times n$ matrix. This set of matrices does not form a group under addition, since it does not fulfill conditions 1 and 3.

5. The set of all real orthogonal $n \times n$ matrices under matrix multiplication forms a group. (1) The product of any two real orthogonal $n \times n$ matrices is a real orthogonal $n \times n$ matrix. (2) Matrix multiplication is associative. (3) The unit $n \times n$ matrix, the identity of matrix multiplication, is real and orthogonal. (4) The inverse of any real orthogonal $n \times n$ matrix is its matrix inverse, which exists and is also a real orthogonal $n \times n$ matrix. This group is non-Abelian for $n > 2$.

6. The set of all displacements (also called translations) in a common direction forms a group with the composition of consecutive displacement. (1) The composition of any two displacements in a common direction is a displacement in the same direction. In fact, the composition of displacement by a cm and displacement by b cm in a common direction is displacement by $(a + b)$ cm in the same direction. We see that the group is Abelian. (2) Composition by consecutive displacement is associative. That might or might not be clear. In Chapter 4 we see that composition of transformations by consecutive transformation is always associative. (3) The identity transformation,

the nonaction of doing nothing (just wonderful for lazy people), is a member of the set, since the null displacement, displacement by 0 cm, is the identity transformation. The identity transformation is the identity of any group of transformations with consecutive transformation as composition. That should be clear but is discussed in detail in Chapter 4. (4) For any displacement the inverse is also a displacement in the same direction. It is displacement by the same distance, in the same direction, but in the opposite sense. In other words, the inverse of displacement by a cm is displacement by $-a$ cm in the same direction.

7. The set of all rotations about a common axis, with composition of consecutive rotation, forms a group. (1) The composition of any two rotations about the same axis is a rotation about the same axis. Rotation by $a°$ followed by rotation by $b°$ about the same axis results in total rotation by $(a + b)°$ about the same axis. The group is clearly Abelian. (2) Composition by consecutive rotation is associative. That is discussed in Chapter 4. (3) The identity transformation is a member of the set, since the null rotation, rotation by $0°$, is the identity transformation. (4) The inverse of any rotation is

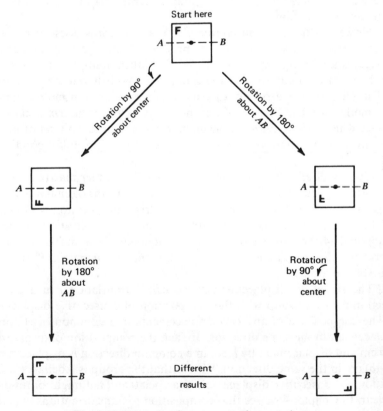

Fig. 2.10. Example of noncommuting rotations.

also a rotation about the same axis. For a rotation by $a°$ the inverse might be expressed as rotation by $-a°$ about the same axis or as rotation by $(360 - a)°$ about the same axis.

8. The set of all rotations about a point, consisting of all rotations about all axes through a point, forms a group under composition of consecutive rotation. (1) Closure is far from obvious, and we do not show it here. (2) Associativity is dicussed in Chapter 4 for transformations in general. (3) The identity transformation is a member of the set; it is the null rotation about any axis through the point. (4) The inverse of any rotation by $a°$ about any axis through the point is rotation by $-a°$ about the same axis and is thus also a member of the set. The group is non-Abelian. To show that consider Fig. 2.10, in which rotation by $90°$ about the axis through the center of the square and perpendicular to the plane of the paper and rotation by $180°$ about axis AB through the center of the square are applied to the marked square consecutively in each possible order. The two results are different. Since at least those two rotations, both members of the group, are non-commuting, the group is non-Abelian.

PROBLEMS

1. (a) Prove that in a finite-order group for every element g there exists a positive integer k such that $g^k = e$. (*Hint:* Consider the sequence $g, g^2, g^3, \ldots$ and remember that the order of the group is finite.) The smallest such k is called the *order* (or *period*) of element g. (b) Find the order of each element of the order-6 group D_3.

2. If a, b and ab are group elements of order 2, prove that a and b commute. (See preceding problem.)

3. Prove that each set of matrices forms a group under matrix multiplication by constructing its group table and identifying the order-4 group of which it is a realization:

   (a) $\left\{ \begin{pmatrix} 1 & 0 \\ 0 & 1 \end{pmatrix}, \begin{pmatrix} 0 & 1 \\ -1 & 0 \end{pmatrix}, \begin{pmatrix} -1 & 0 \\ 0 & -1 \end{pmatrix}, \begin{pmatrix} 0 & -1 \\ 1 & 0 \end{pmatrix} \right\};$

   (b) $\left\{ \begin{pmatrix} 1 & 0 \\ 0 & 1 \end{pmatrix}, \begin{pmatrix} 1 & 0 \\ 0 & -1 \end{pmatrix}, \begin{pmatrix} -1 & 0 \\ 0 & 1 \end{pmatrix}, \begin{pmatrix} -1 & 0 \\ 0 & -1 \end{pmatrix} \right\}.$

4. Prove that the identity transformation and rotations by $180°$ about each of three mutually perpendicular, intersecting straight lines with a common point of intersection form an order-4 group under composition of consecutive transformation. Construct the group table. Of which abstract order-4 group is that a realization?

5. Prove that the set of matrices

$$\left\{ \begin{pmatrix} 1 & 0 \\ 0 & 1 \end{pmatrix}, \begin{pmatrix} \omega & 0 \\ 0 & \omega^2 \end{pmatrix}, \begin{pmatrix} \omega^2 & 0 \\ 0 & \omega \end{pmatrix}, \begin{pmatrix} 0 & 1 \\ 1 & 0 \end{pmatrix}, \begin{pmatrix} 0 & \omega^2 \\ \omega & 0 \end{pmatrix}, \begin{pmatrix} 0 & \omega \\ \omega^2 & 0 \end{pmatrix} \right\},$$

where $\omega^3 = 1$ but $\omega \neq 1$, forms a group under matrix multiplication, and prove that it is a realization of the group D_3.

6. The law of composition of the set of functions

$$\left\{ f_1(x) = x, \quad f_2(x) = \frac{1}{1-x}, \quad f_3(x) = \frac{x-1}{x}, \right.$$

$$\left. f_4(x) = \frac{1}{x}, \quad f_5(x) = 1-x, \quad f_6(x) = \frac{x}{x-1} \right\}$$

is substitution. For example,

$$f_4 f_3 = f_4(f_3(x)) = \frac{1}{f_3(x)} = \frac{x}{x-1} = f_6.$$

Prove that the set forms a group under that composition by constructing its group table, and show that it is a realization of the group D_3.

7. Why does not each set form a group under the specified law of composition (check all four conditions)? (a) All odd integers under addition; (b) $\{\frac{1}{2}, \frac{1}{3}, 1, 2, 3\}$ under multiplication; (c) all positive rational numbers, where the composition of a and b is a/b; and (d) all ordinary vectors (3-vectors) under vector product.

8. Prove that each set of numbers forms an infinite-order (Abelian) group under ordinary multiplication: (a) $\{2^k\}$ for integer k; (b) $\{(1+2m)/(1+2n)\}$ for integer m and n; (c) $\{a + b\sqrt{5}\}$ for rational a and b such that $a^2 + b^2 \neq 0$; and (d) $\{\cos\alpha + i\sin\alpha\}$ for real α.

9. Each of the following sets of numbers forms an infinite-order (Abelian) group either under addition or under multiplication. For each set check which is the group composition and pove that the set is indeed a group: (a) all real numbers; (b) all nonzero real numbers; (c) all positive reals; (d) all complex numbers; (e) all nonzero complex numbers; and (f) all unimodular complex numbers (i.e., all complex numbers whose absolute value is 1).

10. (a) Prove that the set of all real polynomials in one variable forms a group under addition. (b) Why does the set of all nonzero real polynomials in one variable not form a group under multiplication?

11. (a) Prove that the set of all unitary $n \times n$ matrices forms a group under matrix multiplication. (b) Prove that the set of all unimodular (determinant = 1) real orthogonal $n \times n$ matrices forms a group under matrix multiplication.

12. Prove that the set of matrices

$$L(v) = (1 - v^2/c^2)^{-1/2} \begin{pmatrix} 1 & -v \\ -v/c^2 & 1 \end{pmatrix}$$

for all v in the range $-c < v < c$, with c a positive constant, forms an Abelian "continuous" group in the sense that

$$L(v_1)L(v_2) = L(v_3),$$

where

$$v_3 = \frac{v_1 + v_2}{1 + v_1 v_2/c^2}.$$

That is the group of Lorentz transformations in the same direction, where c is the speed of light.

13. Prove that the set of all matrices of the form $\begin{pmatrix} a & -b \\ b & a \end{pmatrix}$ with a and b real forms an Abelian group under matrix multiplication as long as a and b satisfy a certain condition. What is that condition?

2.2. Mapping

At this point I think it is worthwhile to devote a short discussion to the concept of *mapping*, because we make much use of mappings in the rest of the book. A mapping is a correspondence made from one set to another (or to the same set). (The sets may or may not be groups.) A mapping from set A to set B puts every element of A in correspondence with some element of B, as in Fig. 2.11. A mapping is denoted $A \to B$. If the mapping is given a name, say the mapping M from A to B, it is denoted $A \overset{M}{\to} B$ or $M: A \to B$.

An element of B that is in correspondence with an element of A is called the *image* of the element of A. The element of A is called an *object* of the element of B. A mapping $A \to B$ such that an element of B may be the image of more than one element of A is called a *many-to-one* mapping. If, however, different elements of A are always in correspondence with different elements of B, i.e., if every image in B is the image of a unique object in A, the mapping is called *one-to-one*. (A one-to-one mapping is also called an *injective* mapping.) Figure 2.12 illustrates a one-to-one mapping.

If not all elements of B are necessarily images under the mapping, it is called a mapping of A *into* B. Figures 2.11 and 2.12 illustrate such mappings. If, however, every element of B is the image of at least one element of A, the

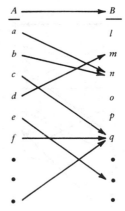

Fig. 2.11. Many-to-one mapping of A into B.

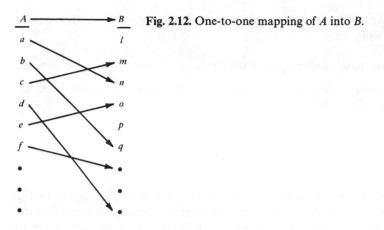

Fig. 2.12. One-to-one mapping of A into B.

mapping is from A *onto* B, as in Fig. 2.13. (An onto mapping is also called a *surjective* mapping.)

A mapping may also be denoted in terms of the set elements, as $a \to b$ or $a \overset{M}{\to} b$, where a represents elements of A and b represents elements of B. We use that notation extensively. Another notation, using the notation commonly used for functions, is $b = M(a)$. We also use this notation, but not in Chapters 2 and 3.

A mapping may be from a set to itself, i.e., A and B may be the same set. Familiar examples of such mappings are real functions,

$$x \to y = f(x),$$

where the set is the set of all real numbers. The function $y = x^2$ is a many-to-one, actually two-to-one except for $y = 0$, mapping of the reals into the reals (only the nonnegative numbers are images). The function $y = e^x$ is a one-to-one mapping of the reals into the reals (only positive numbers are images). The function $y = \sin x$ is an infinity-to-one mapping of the reals into the reals

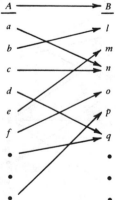

Fig. 2.13. Many-to-one mapping of A onto B.

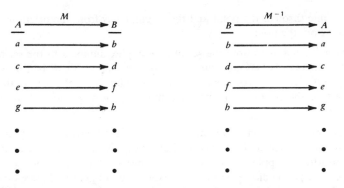

Fig. 2.14. One-to-one mapping M of A onto B and its inverse M^{-1}.

(only numbers whose absolute values are not larger than 1 are images). The function $y = 3x - 7$ is a one-to-one mapping from the reals onto the reals.

A one-to-one mapping from A onto B may be inverted, simply by reversing the sense of the correspondence arrows, to obtain a one-to-one mapping from B onto A. The latter mapping is called the *inverse* of the former. All elements of B are objects under the inverse mapping, because all elements of B are images for the direct mapping. The inverse mapping is one-to-one, since the direct mapping is one-to-one. And the inverse mapping is onto, because all elements of A are objects under the direct mapping. Thus the inverse mapping is indeed a one-to-one mapping from B onto A, as in Fig. 2.14. It inverse is the direct mapping itself. If the direct mapping is denoted M, its inverse is denoted M^{-1}, so that $A \overset{M}{\to} B$, $B \overset{M^{-1}}{\longrightarrow} A$, $a \overset{M}{\to} b$, $b \overset{M^{-1}}{\longrightarrow} a$, $b = M(a)$, $a = M^{-1}(b)$. Note that only a mapping that is both one-to-one and onto has an inverse; none of the other kinds of mapping just discussed is invertible. (A mapping that is both one-to-one and onto, i.e., is both injective and surjective, is called *bijective*. Thus bijectiveness and invertibility imply one another.)

As an example of an inverse mapping, taking the one-to-one mapping

$$x \to y = 3x - 7$$

from the reals onto the reals, the inverse mapping is

$$y \to x = (y + 7)/3.$$

PROBLEMS

1. Each of the following is a mapping of the set of all points of a plane to itself. Prove that each mapping is one-to-one and onto so that it has an inverse. Find the inverse. (a) The image of any point is found by moving from it 3 units in the x direction and 2 units in the y direction; (b) the image of any point is found by moving from it 7° counterclockwise around the circle through the point with center at the origin; and (c) the image of any point is found by moving from it along the

straight line through the point and the origin, away from the origin to twice the distance from the origin.

2. Consider the mapping from the set of all complex numbers to the set of all real numbers: $z \to |z|$. Is that mapping one-to-one or many-to-one? Is it into or onto? Is it invertible? If we considered it as a mapping from the set of all complex numbers to the set of all nonnegative numbers, how would you answer those questions?

3. For each of the following mappings answer the questions: Is the mapping one-to-one or many-to-one, and, if the latter, how many? Is the mapping into or onto? Is it an invertible mapping? If so, what is the inverse mapping? If not, why not? (a) Mapping by perpendicular projection of the set of all points of three-dimensional space to the set of all points of a plane in that space, i.e., the image in the plane of any point in space is found by dropping a perpendicular from the point to the plane; (b) mapping by perpendicular projection of the set of all points of a straight line to the set of all points of a plane that is not perpendicular to the line, i.e., the image of any point on the line is found by dropping a perpendicular from the point to the plane; and (c) mapping by perpendicular projection of the set of all points of a straight line to the set of all points of another straight line that is not orthogonal to the first line, i.e., the image of any point on the first line is found by dropping a perpendicular from the point to the second line.

4. For each of the following functions $y = f(x)$ answer the questions: Does

$$x \to y = f(x)$$

define a mapping $R \to R'$, where both R and R' denote the set of all real numbers? If not, can R be reduced (e.g., to the set of all positive numbers) to make a mapping? Is the mapping one-to-one or many-to-one (how many?)? If it is many-to-one, can R be reduced (further) to make the mapping one-to-one? Is the mapping into or onto? If it is into, reduce R' to make the mapping onto. If you have obtained a one-to-one mapping that is onto, it is invertible. What is the corresponding inverse function? (a) $y = x^2$; (b) $y = e^x$; (c) $y = \sin x$; (d) $y = \tan x$; (e) $y = +\sqrt{x}$; (f) $y = \ln x$; and (g) $y = 2x + \sqrt{5}$.

5. Consider a circle, a straight line tangent to it, and a point P located anywhere in the plane of the circle except on the straight line. Denote the set of all points on the circle by C and the set of all points on the line by L. We define the correspondence between points of C and points of L: A point of C and a point of L are in correspondence if and only if they are collinear with P, i.e., if the straight line through them passes also through P. Consider that correspondence for each of the following locations of P and answer the questions: does the correspondence define a mapping $C \to L$ or $L \to C$ or no such mapping at all? If it defines such a mapping, is it one-to-one, many-to-one (how many?), into, onto? Is it an invertible mapping? (a) The line separates P and the circle; (b) P is on the same side of the line as the circle, but outside the circle; (c) P is inside the circle; (d) P is on the circle, but not diametrically opposite the point of tangency; (e) P is on the circle, diametrically opposite the point of tangency. (*Note:* The question "How many?" might not have the same answer for all images of a mapping. To be able to set up a mapping, you might find it necessary to remove one or more points from C or, alternatively, to add to L the "point at infinity.")

2.3. Isomorphism

Consider a many-to-one or one-to-one mapping of a group G onto another group G', so that every element of G has an image in G' and every element of G' is an image of at least one element of G, as in Fig. 2.15.

Now consider such a mapping that preserves structure. What we mean by this is that, if a and b are any two elements of G with images a' and b', respectively, in G', and if we denote by $(ab)'$ the image in G' of the composition ab in G, then

$$a'b' = (ab)'$$

(where the composition $a'b'$ is, of course, in G'). In other words, the image of a composition is the composition of images. That can also be expressed by the diagram for all a, b in G:

$$a \quad b = c$$
$$\downarrow \quad \downarrow \quad \downarrow$$
$$a' \quad b' = c'.$$

Such a structure-preserving mapping is called a *homomorphism*. We postpone the discussion of homomorphism in general to Section 2.5.

If a homomorphism is one-to-one, i.e., if each element of G' is the image of exactly one element of G so that the mapping can be inverted, it is called an *isomorphism* and is denoted $G \sim G'$. Figure 2.16 shows such a mapping. And the structure-preserving property of an isomorphism can be expressed by the diagram

$$a \quad b = c$$
$$\updownarrow \quad \updownarrow \quad \updownarrow$$
$$a' \quad b' = c'.$$

Isomorphic groups have the same structure. And groups having the same structure, in the sense we used previously, that their group tables can be

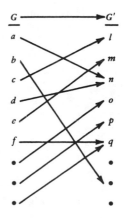

Fig. 2.15. Many-to-one mapping of G onto G'.

$\underline{G} \sim \underline{G'}$ **Fig. 2.16.** Isomorphism mapping of G and G'.

$a \longleftrightarrow a'$

$b \longleftrightarrow b'$

$c \longleftrightarrow c'$

$d \longleftrightarrow d'$

$e \longleftrightarrow e'$

$f \longleftrightarrow f'$

•   •

•   •

•   •

made the same by reordering rows and columns and redenoting elements, are isomorphic. So "isomorphic" makes precise the concept of "having the same structure" and is more general, since it is applicable to infinite-order as well as finite-order groups. Thus all realizations of an abstract group are isomorphic with the abstract group of which they are realizations as well as with each other.

Consider some examples of isomorphism.

1. The abstract group of order 3 C_3 and two of its realizations, by rotations and by complex numbers. See Fig. 2.17.

2. The Abelian infinite-order group of integers under addition is isomorphic with the Abelian infinite-order group of integral powers of 2 (or of any other positive number) under multiplication. The mapping is $n \leftrightarrow 2^n$. The identities are $0 \leftrightarrow 2^0 = 1$. Structure is preserved by the mapping:

$$m + n = (m + n)$$
$$\updownarrow \quad \updownarrow \quad \quad \updownarrow$$
$$2^m \times 2^n = 2^{m+n}.$$

3. The Abelian infinite-order group of all real numbers under addition is

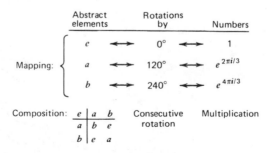

Fig. 2.17. Isomorphism of C_3 and two of its realizations.

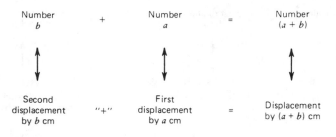

Fig. 2.18. Structure preservation for isomorphism of group of real numbers under addition with group of displacements in common direction (with composition of consecutive displacement, denoted "+").

isomorphic with the Abelian infinite-order group of all displacements in a common direction with the composition of consecutive displacement. The mapping is (number a)$\leftrightarrow$(displacement by a cm). See Fig. 2.18 for structure preservation.

Here and in the following we adopt the convention that consecutive transformations are read from right to left. The reason is made clear in Chapter 4. For the time being we use "+" to denote composition by consecutive transformation.

4. The Abelian infinite-order group of all rotations about a common axis, with composition of consecutive rotation, is isomorphic with the Abelian infinite-order group of all unimodular (determinant = 1) real orthogonal 2×2 matrices under matrix multiplication. Both are isomorphic with the group of unimodular (absolute value = 1) complex numbers under multiplication. Note that the most general 2×2 matrix that is unimodular, real, and orthogonal is $\begin{pmatrix} \cos \alpha & -\sin \alpha \\ \sin \alpha & \cos \alpha \end{pmatrix}$ with α real. The most general unimodular complex number is $e^{i\alpha}$ with α real. The mapping for those isomorphisms is

$$(\text{rotation by angle } \alpha) \leftrightarrow \begin{pmatrix} \cos \alpha & -\sin \alpha \\ \sin \alpha & \cos \alpha \end{pmatrix} \leftrightarrow e^{i\alpha}.$$

The identities are

$$(\text{rotation by angle } 0) \leftrightarrow \begin{pmatrix} 1 & 0 \\ 0 & 1 \end{pmatrix} \leftrightarrow e^0 = 1.$$

For structure preservation see Fig. 2.19.

5. The non-Abelian infinite-order group of all rotations about a common point under consecutive rotation is isomorphic with the non-Abelian infinite-order group of unimodular real orthogonal 3×3 matrices under matrix multiplication. The mapping is (rotation by angle α about axis with direction consines (λ, μ, ν))$\leftrightarrow$(a certain unimodular real orthogonal 3×3 matrix $M(\lambda, \mu, \nu; \alpha)$ uniquely determined by (λ, μ, ν) and α in a way that we do not go into here). See [M46] and [M45]. For structure preservation see Fig. 2.20.

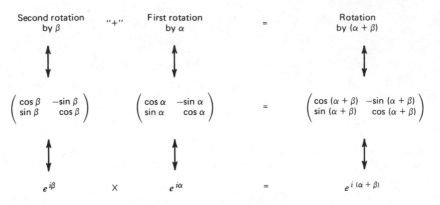

Fig. 2.19. Structure preservation for isomorphism of group of rotations about common axis (with composition of consecutive rotation, denoted "+"), group of unimodular real orthogonal 2 × 2 matrices under matrix multiplication, and group of unimodular complex numbers under multiplication.

PROBLEMS

1. Prove that groups of different finite orders cannot be isomorphic with each other.

2. Prove that in an isomorphism corresponding elements have the same order. (See Problem 1 of Section 2.1.)

3. Prove that a group is Abelian if and only if the correspondence $a \leftrightarrow a^{-1}$, $b \leftrightarrow b^{-1}, \ldots$ for all its elements is an isomorphism (of the group with itself).

4. All the previous examples and problems formulated in terms of realizations are, of course, isomorphism examples and problems. Formulate them rigorously in terms of isomorphism.

5. Prove that the group of all positive numbers under multiplication is isomorphic with the group of all real numbers under addition. (*Hint:* Consider the function $y = e^x$ or its inverse, $x = \ln y$. That isomorphism is the theoretical basis for the use of logarithms and slide rules for calculating products, ratios, powers, and roots. However, by now the pocket calculator has superseded logarithms and slide rules.)

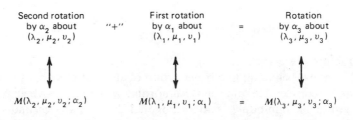

Fig. 2.20. Structure preservation for isomorphism of group of rotations about common point (with composition of consecutive rotation, denoted "+") with group of unimodular real orthogonal 3 × 3 matrices $M(\lambda, \mu, v; \alpha)$ under matrix multiplication.

6. Prove that C_n, the group of rotations about a common axis by $\{0°, 360°/n, 2 \times 360°/n, 3 \times 360°/n, \ldots, (n-1) \times 360°/n\}$, with composition of consecutive rotation, is isomorphic with the group consisting of the n nth roots of 1, $\{1, e^{2\pi i/n}, e^{2 \times 2\pi i/n}, e^{3 \times 2\pi i/n}, \ldots, e^{(n-1) \times 2\pi i/n}\}$, under multiplication. (See the first example in this section.)

7. Prove that the group of matrices of the form $\begin{pmatrix} a & -b \\ b & a \end{pmatrix}$, with a and b real and obeying a certain condition, under matrix multiplication is isomorphic with the group of nonzero complex numbers under multiplication. (See Problem 13 of Section 2.1.)

2.4. Equivalence Relation

At this most appropriate point we take the opportunity to introduce the concept of *equivalence rleation*, of which isomorphism is an example, as we will see. An equivalence relation is defined as any relation, denoted here by $\equiv$, that might hold between the members of pairs of elements of a set and satisfies the following three conditions:

1. *Reflexivity.* Every element of the set has the relation with itself, i.e.,

$$a \equiv a$$

for all a.

2. *Symmetry.* If one element has the relation with another, then the second has it with the first, for all elements of the set. In symbols that is

$$a \equiv b \quad \Leftrightarrow \quad b \equiv a$$

for all a, b. (The arrow $\Rightarrow$ denotes implication in the direction of the arrow. The double arrow $\Leftrightarrow$ denotes implication both ways. Recall that whatever stands at the head of an implication arrow is a necessary condition for whatever stands at the tail. And whatever stands at the tail is a sufficient condition for whatever stands at the head. Whatever stands at one head of a double implication arrow is a necessary and sufficient condition for whatever stands at the other head.)

3. *Transitivity.* If one element has the relation with a second and the second has it with a third, then the first element has the relation with the third, for all elements of the set. Represented symbolically that is

$$a \equiv b, \quad b \equiv c \quad \Rightarrow \quad a \equiv c$$

for all a, b, c.

One somewhat fanciful example of an equivalence relation is friendship. (1) It is a reflexive relation, since everyone is a friend of himself or herself (well, we hope so, anyhow). (2) It is symmetric, since if you are my friend I am your

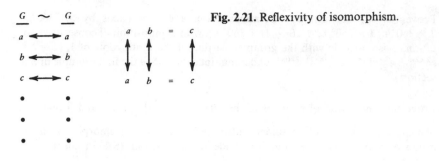

Fig. 2.21. Reflexivity of isomorphism.

friend. (3) And it is transitive, since a friend of a friend is a freiend (again, that is the ideal).

More seriously, the most familiar example of equivalence relation is equality, as denoted by $=$. That hardly needs explaining.

Isomorphism is an equivalence relation. The elements here are groups, members of the set of all groups (and not the elements of the groups!). (1) Isomorphism is reflexive, because every group is isomorphic with itself. The mapping is just every element is its own image. See Fig. 2.21. (2) Isomorphism is symmetric, since the mapping is one-to-one and onto, thus invertible. See Fig. 2.22. (3) Isomorphism is transitive. Figure 2.23 is worth a hundred words.

The congruence relation of modular arithmetic, "$\equiv$ (mod n)," is an equivalence relation.

An example of a relation that is not an equivalence relation is the relation of "lesser than," denoted $<$. It is not reflexive, since no number is lesser than itself. Neither is it symmetric, since, if a is lesser than b, certainly b is not lesser than a. It is, however, transitive; if a is lesser than b and b is lesser than c, then a is lesser than c.

Another example is the relation of inequality, denoted $\neq$. That relation is symmetric but not reflexive or transitive. (If a does not equal b and b does not equal c, nothing is implied about the relation between a and c.)

Intimately related to the concept of equivalence relation is the concept of *equivalence class*, but we postpone presenting it until the next chapter.

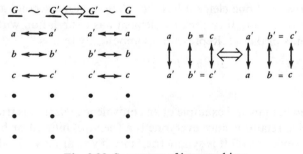

Fig. 2.22. Symmetry of isomorphism.

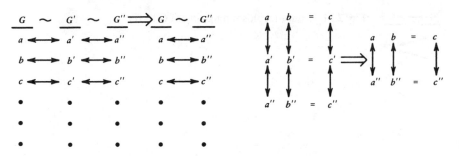

Fig. 2.23. Transitivity of isomorphism.

PROBLEMS

1. You might think that the reflexivity property of an equivalence relation is redundant in that it follows from the other two properties. Show that, if you can. However, a certain assumption is being made. Which? (What about reflexivity for an element that is not equivalent to any other element?)

2. Prove that the congruence relation of modular arithmetic, "$\equiv$ (mod n)," is an equivalence relation.

3. Check whether each of the following relations is an equivalence relation or not. If not, for which of the three defining properties does it fail? (a) The relation $\geq$ among real numbers; (b) having the same absolute value for complex numbers; (c) having the same value of determinant for matrices; (d) parallelism among straight lines in three-dimensional space; (e) orthogonality among straight lines in three-dimensional space; and (f) having at least one root in common for quadratic equations.

2.5. Homomorphism

We now return to the discussion of homomoprhism (of which isomorphism is a special case) in general, which we abandoned in Section 2.3. Such a mapping is not invertible (unless it is an isomorphism and only then). Homomorphism is denoted $G \to G'$, where the mapping is many-to-one from G to G', as in Fig. 2.24. The structure-preserving property of homomorphism can be expressed by the diagram

$$a \quad b = c$$
$$\downarrow \quad \downarrow \quad \downarrow$$
$$a' \quad b' = c'.$$

Although homomorphism preserves structure in the sense just shown, homomorphic groups do not have the same structure, unless they are isomorphic. For example, non-Abelian groups may be homomorphic to Abelian groups (the opposite is impossible, however). Infinite-order groups may be homomorphic to finite-order groups, and groups of different finite orders

$G \longrightarrow G'$ **Fig. 2.24.** Homomorphism mapping of G to G'.

may be homomorphic. Examples are presented shortly. Homomorphism is not an equivalence relation, since it is not symmetric. It is reflexive, however, with isomorphism as a special case of homomorphism. See Fig. 2.21. And it is transitive. See Fig. 2.25.

In a homomorphism $G \to G'$ the set of all elements of G whose images are the identity element of G' is called the *kernel* of the homomorphism. We will return to that in Section 3.1.

Some examples of homomorphism follow.

1. Any group whatsoever is trivially homomorphic to the abstract group of order 1 C_1, as in Fig. 2.26. Structure is preserved:

$$a \ \ b = c$$
$$\downarrow \ \downarrow \ \ \downarrow$$
$$e' \ e' = e'.$$

All of G is the kernel of such a homomorphism. In terms of realizations, we can homomorphically map all the elements of any group to the number 1 (under multiplication) or to the number 0 (under addition).

2. Both abstract groups of order 4 (or any realizations thereof) are homomorphic to the abstract group of order 2 C_2 (or any of its realizations). For C_4 the mapping is shown in Fig. 2.27. For D_2 there are three possible mappings, shown in Fig. 2.28. Structure preservation can be checked with the help of the group tables (Section 2.1). The kernel of the first homomorphism (Fig. 2.27) is $\{e, b\}$. The three possible kernels of the second homomorphism (Fig. 2.28) are $\{e, a\}$, $\{e, b\}$, $\{e, c\}$.

3. The non-Abelian order-6 group D_3 is homomorphic to Abelian C_2. The mapping is shown in Fig. 2.29. Preservation of structure is checked with the group tables (Section 2.1). The kernel of that homomorphism is $\{e, a, b\}$.

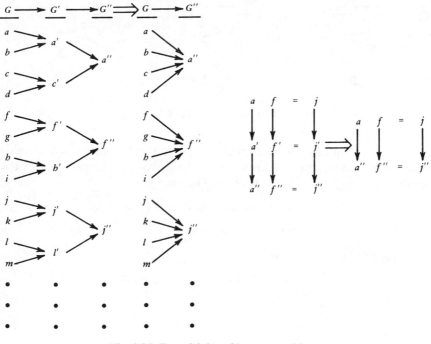

Fig. 2.25. Transitivity of homomorphism.

Fig. 2.26. Trivial homomorphism of any group to order-1 group.

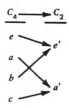

Fig. 2.27. Homomorphism of C_4 to C_2.

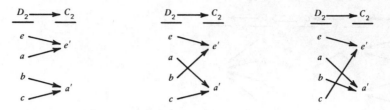

Fig. 2.28. Homomoprhism of D_2 to C_2.

4. The Abelian infinite-order group of nonzero real numbers under multiplication is homomorphic to Abelian C_2 or its realization, for example, by the numbers $\{1, -1\}$ under multiplication. The mapping is

$$\text{any positive number} \rightarrow e \leftrightarrow 1,$$

$$\text{any negative number} \rightarrow a \leftrightarrow -1.$$

Structure preservation is seen as follows:

$$
\begin{array}{ccccc}
\text{positive} & & \text{positive} & & \text{positive} \\
\text{number} & \times & \text{number} & = & \text{number} \\
\downarrow & & \downarrow & & \downarrow \\
e & & e & = & e \\
\updownarrow & & \updownarrow & & \updownarrow \\
1 & \times & 1 & = & 1
\end{array}
$$

$$
\begin{array}{ccccc}
\text{positive} & & \text{negative} & & \text{negative} \\
\text{number} & \times & \text{number} & = & \text{number} \\
\downarrow & & \downarrow & & \downarrow \\
e & & a & = & a \\
\updownarrow & & \updownarrow & & \updownarrow \\
1 & \times & (-1) & = & (-1)
\end{array}
$$

$$
\begin{array}{ccccc}
\text{negative} & & \text{positive} & & \text{negative} \\
\text{number} & \times & \text{number} & = & \text{number} \\
\downarrow & & \downarrow & & \downarrow \\
a & & e & = & a \\
\updownarrow & & \updownarrow & & \updownarrow \\
(-1) & \times & 1 & = & (-1)
\end{array}
$$

$$
\begin{array}{ccccc}
\text{negative} & & \text{negative} & & \text{positive} \\
\text{number} & \times & \text{number} & = & \text{number} \\
\downarrow & & \downarrow & & \downarrow \\
a & & a & = & e \\
\updownarrow & & \updownarrow & & \updownarrow \\
(-1) & \times & (-1) & = & 1
\end{array}
$$

The set of all positive numbers comprises the kernel of that homomorphism.

Fig. 2.29. Homomorphism of D_3 to C_2.

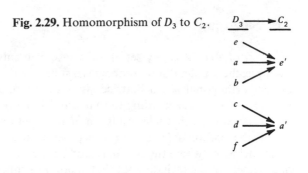

5. The non-Abelian (for $n > 1$) group of all nonsingular complex $n \times n$ matrices under matrix multiplication is homomorphic to the Abelian group of nonzero complex numbers under multiplication. Each matrix A is mapped to the complex number that is the value of its determinant, $A \to |A|$ (so that $A^{-1} \to 1/|A|$). Structure preservation follows from the fact that the determinant of a matrix product is the product of the individual determinants:

$$A \qquad B \; = (AB)$$
$$\downarrow \qquad \downarrow \qquad \downarrow$$
$$|A| \times |B| = |AB|.$$

The unimodular $n \times n$ matrices comprise the kernel of that homomorphism, since they are all the matrices whose image is 1, the identity of number multiplication.

PROBLEMS

1. Prove that an Abelian group cannot be homomorphic to a non-Abelian group.

2. Prove that the group of nonzero real numbers under multiplication is homomorphic to the group of positive numbers under multiplication. What is the kernel?

3. (a) Prove that the group of nonzero complex numbers under multiplication is homomorphic to the group of positive numbers under multiplication. What is the kernel? (b) Find two different homomorphisms of the group of complex numbers under addition to the group of real numbers under addition. What are the kernels?

4. Prove that the group of nonsingular complex $n \times n$ matrices under matrix multiplication is homomorphic to the group of unimodular complex $n \times n$ matrices under matrix multiplication. (*Hint:* Consider the mapping: $A \to A/|A|^{1/n}$.) What is the kernel?

5. Prove that the operation of differentiation maps the group of all polynomials in one variable (under addition) onto itself and prove that this mapping is a homomorphism. What is the kernel?

2.6. Subgroup

A *subset* of a set A is any set all of whose elements are elements of A. For example, the set of natural numbers is a subset of the set of integers, or the set of real orthogonal $n \times n$ matrices is a subset of the set of unitary $n \times n$ matrices. Note that according to that definition any set is a subset of itself. If B is a subset of A, that relation is denoted $B \subset A$ or $A \supset B$, where $\subset$ and $\supset$ denote the relation of *inclusion*. (Sometimes $\subset$ and $\supset$ are used to denote only strict inclusion, where the sets may not be the same, so that there is at least one element in the including set that is not a member of the subset. Inclusion that allows the subset to coincide with the including set is then denoted $\subseteq$ or $\supseteq$. However, we do not use that convention.)

If a subset of the elements of a group G itself forms a group with respect to the composition law of G, it is called a *subgroup* of G. If H is a subgroup of G, the relation is denoted $H \subset G$ or $G \supset H$, using the same notation used for set inclusion.

Every group G has two trivial subgroups: (1) the group of order 1 consisting of the identity element of G; and (2) the group G itself. All the subgroups of a group except the group itself are called *proper subgroups*.

It can be shown, and indeed we show in Section 3.2, that if a finite-order group of order n includes a subgroup of order m, then m is a divisor of n, i.e., $n = ms$ for some integer s. From that theorem it follows that groups of prime order have no proper subgroups except the trivial subgroup of order 1. Thus the groups of orders 2, 3, and 5, which were presented in Section 2.1, can have no nontrivial proper subgroups, as an examination of their group tables can confirm. In any case the group of order 2 is of too low an order to have nontrivial proper subgroups.

If G_2 is a subgroup of G_1 and G_3 a subgroup of G_2, then clearly G_3 is also a subgroup of G_1.

Consider the following examples of subgroups.

1. In the order-4 group C_4 (rotations by $\{0°, 90°, 180°, 270°\}$) the elements $\{e, b\}$ (rotations by $\{0°, 180°\}$) form an order-2 subgroup C_2. That is the only subgroup of C_4.

2. The order-4 group D_2 includes three order-2 subgroups: $\{e, a\}$, $\{e, b\}$, $\{e, c\}$. According to the above theorem the order-4 groups cannot have order-3 subgroups, so we have here all their nontrivial proper subgroups.

3. An order-6 group can have nontrivial proper subgroups of orders 2 and 3 only. In fact, the non-Abelian order-6 group D_3 includes four such subgroups, one of order 3 and three of order 2: $\{e, a, b\}$, $\{e, c\}$, $\{e, d\}$, $\{e, f\}$.

4. The group of nonzero real numbers under multiplication includes various subgroups. Among those are the positive real numbers, the nonzero rational numbers, the positive rational numbers, and the integral powers of a fixed real number.

5. The group of nonsingular complex $n \times n$ matrices under matrix multi-

plication has various ($n \times n$)-matrix subgroups. Among them are the unitary matrices, the nonsingular real matrices, the complex orthogonal matrices, the real orthogonal matrices, and the unimodular matrices.

PROBLEMS

1. Prove that a necessary and sufficient condition for a subset H of a group G to be a subgroup of G is that for all elements h_1 and h_2 of H, $h_1 h_2^{-1}$ is an element of H.

2. The *intersection* of sets H_1 and H_2, denoted $H_1 \cap H_2$, is the set of all elements each of which is a member of both H_1 and H_2. Prove that if H_1 and H_2 are subgroups of group G, then $H_1 \cap H_2$ is also a subgroup of G.

3. Every group contains at least one element that commutes with all elements of the group, i.e., an element a such that $ag = ga$ for all elements g of the group. The identity e is such an element, and there may be more. The set of all such elements of a group is called the *center* of the group. Prove that the center of a group is a subgroup. What is the center of an Abelian group?

4. An Abelian group cannot have a non-Abelian subgroup, but a non-Abelian group can have an Abelian subgroup. Find examples of the latter.

5. The inclusion relation between groups is not an equivalence relation. But which of the properties of an equivalence relation does it fulfill?

6. The group table of the order-8 group denoted D_4 is shown in Fig. 2.30. Find and identify all proper subgroups of D_4.

7. The group table of the order-8 group denoted Q_4 (and called the *quaternion group*) is shown in Fig. 2.31. Find and identify all proper subgroups of Q_4.

8. (a) Prove that the group of integers under addition is isomorphic with the group of even integers under addition. (b) Prove that the latter is a proper subgroup of the former. (c) If a group is isomorphic with a proper subgroup of itself, what can be stated about the order of the group?

9. Prove that each of the following sets of numbers forms a subgroup of the group of all nonzero real numbers under multiplication: (a) all positive real numbers; (b) all nonzero rational numbers; (c) all positive rational numbers; and (d) the integral powers of a fixed real number.

e	a	b	c	d	f	g	h
a	b	c	e	f	g	h	d
b	c	e	a	g	h	d	f
c	e	a	b	h	d	f	g
d	h	g	f	e	c	b	a
f	d	h	g	a	e	c	b
g	f	d	h	b	a	e	c
h	g	f	d	c	b	a	e

Fig. 2.30. Group table of D_4.

e	a	b	c	d	f	g	h
a	b	c	e	f	g	h	d
b	c	e	a	g	h	d	f
c	e	a	b	h	d	f	g
d	h	g	f	b	a	e	c
f	d	h	g	c	b	a	e
g	f	d	h	e	c	b	a
h	g	f	d	a	e	c	b

Fig. 2.31. Group table of Q_4.

10. Check if each of the following sets is a subgroup of the group of all real numbers under addition, the group of all nonzero real numbers under multiplication, or neither: (a) all integral multiples of a fixed real number; (b) $\{a^b\}$, where a is a fixed positive real number and b runs over the preceding set; (c) all nonzero pure imaginary complex numbers; (d) all irrational numbers together with the number 1; and (e) all irrational numbers together with the number 0.

11. Prove that each of the following sets of $n \times n$ matrices is a subgroup of the group of all nonsingular complex $n \times n$ matrices under matrix multiplication: (a) all complex orthogonal matrices; (b) all nonsingular diagonal matrices; (c) all nonsingular real matrices; (d) all nonzero complex scalar matrices. Are any of those sets subgroups of others of those?

12. Prove that the following groups of $n \times n$ matrices under matrix multiplication form an inclusion chain, where each group is a proper subgroup of each of the preceding groups: (a) all nonsingular complex matrices; (b) all unitary matrices; (c) all real orthogonal matrices; (d) all unimodular real orthogonal matrices; and (e) all diagonal unimodular real orthogonal matrices.

13. Is each of the following sets a subgroup of the group of all polynomials in one variable under addition? (a) All polynomials of degree less than 2; (b) all polynomials of even degree; (c) all polynomials containing only even powers of the variable; (d) all polynomials of odd degree; (e) all polynomials containing only odd powers of the variable; and (f) all polynomials with rational coefficients.

2.7. Summary of Chapter Two

In this chapter we started an introduction to group theory, the mathematical language of symmetry, and to related topics. To summarize, we list the more important concepts presented in each section.

In Section 2.1: group, group element, group order, group composition, closure of composition, associativity of composition, identity element, inverse element, commuting elements, Abelian (commutative) group, group structure, group table, realization of group, order (or period) of element.

In Section 2.2: mapping, object of mapping, image of mapping, many-to-one mapping, one-to-one mapping, into mapping, onto mapping, inverse mapping.

In Section 2.3: homomorphism, isomorphism.

In Section 2.4: equivalence relation, reflexivity of relation, symmetry of relation, transitivity of relation.

In Section 2.5: homomorphism, kernel of homomorphism.

In Section 2.6: subgroup, proper subgroup.

CHAPTER THREE

Group Theory Continued

We continue our study of group theory and other mathematical concepts from where we left off at the end of the previous chapter. The chapter break has no essential significance and is for purely technical reasons.

3.1. Conjugacy, Equivalence Class, Invariant Subgroup, Kernel

If for some pair of elements a and b of group G there exists a (not necessarily unique) element u of G such that

$$u^{-1}au = b,$$

then a and b are called *conjugate elements* in G. This conjugacy is denoted

$$a \equiv b.$$

It is an equivalence relation among group elements. (1) Conjugacy is reflexive, since every element a is conjugate with itself, $a \equiv a$ for all a in G. That is due to the fact that $e^{-1}ae = a$. (2) Conjugacy is symmetric, since for all a and b in G if $a \equiv b$, then $b \equiv a$, because if there exists an element u of G such that $u^{-1}au = b$, then $v^{-1}bv = a$ with $v = u^{-1}$. (3) And finally, conjugacy is transitive, since for all a, b, c in G if $a \equiv b$ and $b \equiv c$, then $a \equiv c$. If there exist elements u and v of G such that $u^{-1}au = b$ and $v^{-1}bv = c$, then $w^{-1}aw = c$ with $w = uv$.

Note that more than two elements might be conjugate with each other. The identity element cannot be conjugate with any other element. In an Abelian group there are no conjugate elements.

As an example, we find the conjugate elements of the order-6 group D_3 (Fig. 2.9). First we final inverse: $a^{-1} = b$, $b^{-1} = a$, $c^{-1} = c$, $d^{-1} = d$, $f^{-1} = f$. We now conjugate each element of the group by all the others to find conjugates, for example:

$$c^{-1}ac = cac = b,$$

$$d^{-1}cd = dcd = f,$$

$$f^{-1}cf = fcf = d.$$

In that way we finally obtain the breakdown into conjugate elements:

$$e, \qquad a \equiv b, \qquad c \equiv d \equiv f.$$

A subset of elements of a group that consists of a complete set of mutually conjugate elements is called a *conjugacy class* of the group. The identity element, since it is not conjugate with any other element, is always in a conjugacy class of its own. In an Abelian group every element is in a conjugacy class of its own.

For example, the conjugacy classes of D_3 are $\{e\}$, $\{a, b\}$, $\{c, d, f\}$, as we just found.

At this point we introduce the concept of *equivalence class*, of which conjugacy class is a special case. If we have a set of elements for which an equivalence relation is defined, any subset that contains a complete set of mutually equivalent elements is called an equivalence class. It is easily shown that two different equivalence classes cannot have common elements. Thus an equivalence relation brings about a decomposition of the set for which it is defined, such that every element of the set is a member of one and only one equivalence class.

Two examples of equivalence class already noted are conjugacy classes, where the equivalence relation is conjugacy between group elements, and classes of groups with the same structure, where the equivalence relation is isomorphism of groups in the set of all groups.

Conversely, any decomposition of a set into subsets such that every element of the set is contained in one and only one subset defines an equivalence relation. The subsets may be declared, by fiat, equivalence classes, and the corresponding equivalence relation is simply that two elements are equivalent if and only if they belong to the same subset.

Two trivial equivalence class decompositions for any set of elements are the following: The most exclusive equivalence relation, that every element is equivalent only to itself and to no other, decomposes the set into as many equivalence classes as there are elements in the set, since every equivalence class contains but a single element. And, on the other hand, the most inclusive equivalence relation, that all elements of the set are equivalent to each other, makes the whole set a single equivalence class in itself.

Returning now to conjugation, conjugation by a single element has an interesting and useful property. Let a, a', b, b', c, c' be elements of group G such that $a \equiv a'$, $b \equiv b'$, $c \equiv c'$ with respect to conjugation by the same element u of G:

$$u^{-1} \begin{bmatrix} a \\ b \\ c \end{bmatrix} u = \begin{bmatrix} a' \\ b' \\ c' \end{bmatrix}.$$

Now assume $ab = c$. Compose each side of that relation with u^{-1} on the left and with u on the right. The right-hand side becomes $u^{-1}cu = c'$. With some clearly indicated manipulations the left-hand side becomes

$$u^{-1}abu = u^{-1}aebu$$
$$= u^{-1}a(uu^{-1})bu$$
$$= (u^{-1}au)(u^{-1}bu)$$
$$= a'b'.$$

Thus $a'b' = c'$, and intragroup relations are invariant under conjugation by the same element. Note the analogy with invariance of algebraic relations among matrices under similarity transformations.

For example, in group D_3 $bd = f$. Under conjugation by c

$$c^{-1}\begin{bmatrix} b \\ d \\ f \end{bmatrix} c = c\begin{bmatrix} b \\ d \\ f \end{bmatrix} c = \begin{bmatrix} a \\ f \\ d \end{bmatrix},$$

and indeed $af = d$.

If H is a subset (not necessarily a subgroup) of group G, $g^{-1}Hg$ for some g in G denotes the subset of G consisting of elements $g^{-1}hg$, where h runs over all elements of H, i.e., the subset conjugate with H by g. If G is Abelian, $g^{-1}Hg = H$ for all g in G.

If H is a subgroup of G, then $g^{-1}Hg$ for any g in G is also a subgroup of G. (1) Since H is closed under the composition of G, so is $g^{-1}Hg$, since, as shown above, intragroup relations are invariant under conjugation by the same element (here g). (2) Associativity holds, since it holds in G. (3) $g^{-1}Hg$ contains the identity, since H contains it and $g^{-1}eg = e$. (4) Any element of $g^{-1}Hg$ is expressible as $g^{-1}hg$, where h is some element of H. Its inverse is $(g^{-1}hg)^{-1} = g^{-1}h^{-1}g$, which is an element of $g^{-1}Hg$, since h^{-1} is an element of H. Thus $g^{-1}Hg$ contains the inverses of all its elements. So $g^{-1}Hg$ is indeed a group if H is. And, again since intragroup relations are invariant under conjugation by the same element, conjugate subgroups are isomorphic.

For example, D_3 contains the subgroup $\{e, c\}$. The conjugate subgroup by a is

$$a^{-1}\{e, c\}a = b\{e, c\}a = \{e, f\}.$$

Both subgroups are isomorphic, since they are both of order 2, and there is only one abstract order-2 group, C_2.

If H is a subgroup of G and $g^{-1}Hg = H$ for all g in G, i.e., if H is conjugate only with itself, H is called an *invariant subgroup* (also *normal subgroup*).

As an example, the subgroup $\{e, a, b\}$ of D_3 is an invariant subgroup. Earlier in this section we found that elements a and b are conjugate with each other and with no other element, while the identity e is conjugate only with itself. Thus

$$g^{-1}\{e, a, b\}g = \{e, a, b\}$$

for all elements g of D_3 and the subgroup is proved invariant.

Note that the trivial subgroups $\{e\}$ and G of any group G are invariant. In an Abelian group all subgroups are invariant.

Now consider the homomorphism $G \rightarrow G'$. The subset of elements of G that is mapped to the identity element e' of G' is called the *kernel* of the homomorphism, as we mentioned in Section 2.5. A kernel is usually denoted K. Thus K consists of all g in G for which $g \rightarrow e'$. For any homomorphism $G \rightarrow G'$ the kernel K is an invariant subgroup of G, as we immediately prove.

First we prove that K is subgroup of G.

1. Let a and b be any pair of elements of K, so that $a, b \rightarrow e'$. Denote the image of their composition c ($= ab$) by c'. Preservation of structure requires

$$
\begin{array}{ccc}
a & b & = c \\
\downarrow & \downarrow & \downarrow \\
e' & e' & = c'.
\end{array}
$$

But $e'e' = e'$, so $c' = e', c \rightarrow e'$, and $c \in K$. Thus K is closed with respect to the composition of G.

2. Associativity holds in K, since it holds in G.

3. Is e an element of K? The identity e obeys $ea = a$ for all a in G. Let $e \rightarrow x'$ and $a \rightarrow a'$. Preservation of structure requires

$$
\begin{array}{ccc}
e & a & = a \\
\downarrow & \downarrow & \downarrow \\
x' & a' & = a'.
\end{array}
$$

Thus $x'a' = a'$ for all a' in G', while means that $x' = e'$, since the identity is unique, so that $e \rightarrow e'$, and e is indeed an element of K.

4. Let a be any element of K, so that $a \rightarrow e'$, and denote the image of its inverse by $(a^{-1})'$. By preservation of structure

$$
\begin{array}{ccc}
a & a^{-1} & = e \\
\downarrow & \downarrow & \downarrow \\
e' & (a^{-1})' & = e'.
\end{array}
$$

Thus $(a^{-1})' = e'$, and a^{-1} is an element of K. So K contains the inverses of all its elements. That proves that K is a group.

We now prove that K is invariant. Let a be any element of K, so that $a \rightarrow e'$, and from the conjugation $g^{-1}ag = b$ for all elements g of G. Denote by b' the image of b, and denote by g' the image of g and by $(g^{-1})'$ the image of g^{-1}. Preservation of structure gives us $(g^{-1})' = g'^{-1}$ and

$$
\begin{array}{cccc}
g^{-1} & a & g & = b \\
\downarrow & \downarrow & \downarrow & \downarrow \\
g'^{-1} & e' & g' & = b'
\end{array}
$$

from which it follows that $b' = e'$ and $b \in K$. Thus $g^{-1}ag$ is an element of K

for all a in K and all g in G, i.e., $g^{-1}Kg = K$ for all g in G, and K is invariant as claimed.

If the kernel of a homomorphism $G \to G'$ consists only of the identity element e, the mapping is, as will be seen in Section 3.4, one-to-one, and the homomorphism is actually an isomorphism. If the kernel is the group G itself, the homomorphism is the trivial one $g \to e'$ for all g in G, where G' is the order-1 group.

Consider some examples of homomorphism kernels as invariant subgroups.

1. For the homomorphism $D_3 \to C_2$, presented as an example in Section 2.5, the kernel is $\{e, a, b\}$. It was shown in the preceding example in the present section that it is an invariant subgroup.

2. Consider the homomorphism, presented as an example in Section 2.5, of the group of nonsingular complex $n \times n$ matrices under matrix multiplication to the group of nonzero complex numbers under multiplication, where each matrix is mapped to the complex number that is the value of its determinant, $A \to |A|$. The kernel of that homomorphism is the subset of all unimodular $n \times n$ matrices. That subset is a subgroup, as mentioned in Section 2.6. *Proof*: (1) The product of unimodular $n \times n$ metrices is a unimodular $n \times n$ martrix ($|AB| = |A||B|$), so the subset is closed under matrix multiplication. (2) Matrix multiplication is associative. (3) the unit $n \times n$ matrix, the identity of matrix multiplication, is unimodular. (4) The inverse of any unimodular $n \times n$ matrix exists and is unimodular ($|A^{-1}| = 1/|A|$). So the kernel is confirmed to be a subgroup (as was proved in general in this section). Now check if it is an invariant subgroup (which it must be, as was proved in general). Let A be any nonsingular $n \times n$ matrix and U any unimodular $n \times n$ matrix. Form the conjugate element of U by A, $A^{-1}UA$ (which is just the similarly transformation of U by A). The value of its determinant is

$$|A^{-1}UA| = |A^{-1}||U||A|$$
$$= (1/|A|)|U||A|$$
$$= |U|$$
$$= 1$$

for all unimodular U and all nonsingular A, so $A^{-1}UA$ is always unimodular and an element of the kernel, which is thus confirmed to be an invariant subgroup.

3. As seen in Section 2.5, the group of all nonzero real numbers under multiplication is homomorphic to the group of numbers $\{1, -1\}$ under multiplication with mapping

positive number $\to 1$,

negative number $\to -1$.

The kernel is the subset of all positive real numbers, which is indeed a group under multiplication. It is trivially invariant, since multiplication of numbers is commutative.

PROBLEMS

1. Prove the statement made in the text, that two different equivalence classes cannot have common elements, i.e., that two equivalence classes either are disjoint (have no element in common) or are one and the same.

2. If H is an invariant subgroup of G and h is any element of H, prove that the conjugacy class of h in G is a subset of H.

3. Prove that the center of a group is an invariant subgroup. (See Problem 3 of Section 2.6.)

4. For each of the equivalence relations of Problems 2 and 3 of Section 2.4 what are the equivalence classes?

5. For the order-8 group D_4 find all conjugate elements. Which of its proper subgroups are invariant? (See Problem 6 of Section 2.6.)

6. For the order-8 group Q_4 find all conjugate elements. Which of its proper subgroups are invariant? (See Problem 7 of Section 2.6.)

7. For the specific case of the homomorphism of the group of nonsingular complex $n \times n$ matrices to the group of unimodular complex $n \times n$ matrices show that the kernel is an invariant subgroup. (See Problem 4 of Section 2.5.)

3.2. Coset Decomposition

Let H by any proper subgroup of group G. Form the set aH, where a is any element of G that is not a member of H. That is the set of elements ah, where h runs over all the elements of H. Such a set aH is called a *left coset* of H in G. (Similarly, Ha is called a *right coset*.) Clearly, aH is not a subgroup of G, since it does not contain the identity element e. (For aH to contain e, H would have to contain a^{-1}, which it does not, since it is a group and does not contain a.) If H is of order m, then aH also contains m elements. *Proof*: aH certainly contains no more than m elements and could contain less only if $ah_1 = ah_2$ for some distinct pair of elements h_1 and h_2 of H. But $ah_1 = ah_2$ implies $h_1 = h_2$ by composition with a^{-1} on the left, which is a contradiction. Therefore, aH contains exactly m elements. In addition, no element of aH is also an element of H. Otherwise we would have $ah_1 = h_2$ for some pair of elements h_1 and h_2 of H, which implies $a = h_2 h_1^{-1}$ by composition with h_1^{-1} on the right. $a = h_2 h_1^{-1}$ means that a is a member of H, contrary to assumption. So H and aH have no element in common.

The *union* of a number of given sets is the set of all distinct elements of any of the given sets. The symbol $\cup$ is used to denote union. Form the union of

H and aH, $H \cup aH$. The set $H \cup aH$ might or might not exhaust all the elements of G. If it does not, take any element b of G that is not a member of the set $H \cup aH$, i.e., b is an element neither of H nor of aH, and form the left coset bH. By reasoning similar to the above bH has exactly m elements and has no element in common with H or aH. Now form the union $H \cup aH \cup bH$. If this set does not exhaust G, continue forming left cosets in the same manner. If G is of finite order n, we must eventually exhaust it by that procedure:

$$G = H \cup aH \cup bH \cup \cdots \cup kH.$$

Since H and each of its cosets have m elements per set and the sets have no element in common, we obtain the result that $n = ms$ for some integer s, as mentioned in Section 2.6. That result is Lagrange's theorem (Joseph Louis Lagrange, French mathematician and astronomer, 1736–1813). (The whole proof could be performed just as well with right cosets.)

The decomposition of a group G into cosets of any subgroup will always be the same for left cosets as for right cosets if G if Abelian. If G is non-Abelian, the decompositions might be different, although not necessarily. The decomposition of G into left (or right) cosets of subgroup H is unique. By that we mean that, if we form a left coset lH of H with any element l of G that is not a member of H and is also not among the elements a, b, ..., k used in the decomposition

$$G = H \cup aH \cup bH \cup \cdots \cup kH,$$

the coset lH will necessarily be one of the cosets $aH, bH, \ldots, kH$. To see that, note that l must be a member of one of those cosets, say aH. Since all elements of aH are of the form ah, where h runs over all elements of H, then $l = ah_1$ for some element h_1 of H. Thus $a = lh_1^{-1}$ (by composition with h_1^{-1} on the right), and all elements of aH are of the form

$$ah = (lh_1^{-1})h = l(h_1^{-1}h),$$

where h, and consequently also $h_1^{-1}h$, runs over all elements of H. Therefore $aH = lH$. So the same decomposition of G into left (or right) cosets of subgroup H is obtained whatever elements are used to form the cosets.

Examples of decomposition into cosets (refer to Section 2.1 for group tables of the groups involved) follow.

1. The order-4 group C_4 has a single nontrivial proper subgroup, $H = \{e, b\}$. Its left coset formed with a is

$$aH = \{ae, ab\} = \{a, c\}.$$

Its left coset formed with c is

$$cH = \{ce, cb\} = \{c, a\} = aH.$$

So the unique decomposition into cosets of H is

$$G = \{e, b\} \cup \{a, c\}.$$

Since the group is Abelian, decomposition into right cosets of H is the same, which is obvious in this example, since there is only a single possibility for any coset of H.

2. The order-4 group D_2 has three nontrivial proper subgroups: $\{e, a\}$, $\{e, b\}$, $\{e, c\}$. So we have three different decompositions into left cosets:

$$H = \{e, a\}, bH = cH = \{b, c\}, G = \{e, a\} \cup \{b, c\};$$

$$H = \{e, b\}, cH = aH = \{c, a\}, G = \{e, b\} \cup \{c, a\};$$

$$H = \{e, c\}, aH = bH = \{a, b\}, G = \{e, c\} \cup \{a, b\}.$$

Right and left cosets of the same subgroup are the same in this example, since D_2 is Abelian.

3. The non-Abelian order-6 group D_3 has the nontrivial proper subgroups $\{e, a, b\}, \{e, c\}, \{e, d\}, \{e, f\}$. So we have four decompositions into left cosets:

$$H = \{e, a, b\}, cH = \{c, d, f\}, G = \{e, a, b\} \cup \{c, d, f\};$$

$$H = \{e, c\}, aH = \{a, f\}, bH = \{b, d\}, G = \{e, c\} \cup \{a, f\} \cup \{b, d\};$$

$$H = \{e, d\}, aH = \{a, c\}, bH = \{b, f\}, G = \{e, d\} \cup \{a, c\} \cup \{b, f\};$$

$$H = \{e, f\}, aH = \{a, d\}, bH = \{b, c\}, G = \{e, f\} \cup \{a, d\} \cup \{b, c\}.$$

Since the group is non-Abelian, its decompositions into right cosets might be different from its decompositions into left cosets. Its four decompositions into right cosets are

$$H = \{e, a, b\}, Hc = \{c, f, d\}, G = \{e, a, b\} \cup \{c, f, d\};$$

$$H = \{e, c\}, Ha = \{a, d\}, Hb = \{b, f\}, G = \{e, c\} \cup \{a, d\} \cup \{b, f\};$$

$$H = \{e, d\}, Ha = \{a, f\}, Hb = \{b, c\}, G = \{e, d\} \cup \{a, f\} \cup \{b, c\};$$

$$H = \{e, f\}, Ha = \{a, c\}, Hb = \{b, d\}, G = \{e, f\} \cup \{a, c\} \cup \{b, d\}.$$

Thus decomposition into cosets of subgroup $\{e, a, b\}$ is the same for left and right, whereas decomposition into cosets of any of the order-2 subgroups is different for left and right.

Let H be an invariant proper subgroup of group G, so that $g^{-1}Hg = H$ for all g in G. By composition with g on the left, that relation can be put in the form $Hg = gH$ for all g in G, with the result that left and right cosets of any invariant subgroup are the same. (Be careful! $Hg = gH$ for all g in G does not mean that all elements of G commute with all elements of H, but rather that the set of elements hg, where h runs over all elements of H, is the same as the set of elements gh, where h runs over all elements of H, for all g in G.) Thus decomposition of a group into cosets of an invariant subgroup is the same whether left or right cosets are used. We saw that in the example of decomposition of D_3 into cosets of its invariant subgroup $\{e, a, b\}$. (In an Abelian group all subgroups are invariant, so left-coset decomposition and right-coset decomposition are the same for any subgroup.)

PROBLEMS

1. If H is a subgroup of G, prove that a necessary and sufficient condition for elements a and b of G to belong to the same right coset of H is $ab^{-1} \in H$ or, equivalently, $ba^{-1} \in H$. Find corresponding conditions for a and b to belong to the same left coset of H.

2. If H is a subgroup of G, we have seen that the left (or right) cosets of H are disjoint and exhaust G and are thus equivalence classes. Find a succinct formulation of the corresponding equivalence relation.

3. If H is a subgroup of G and

$$G = H \cup aH \cup bH \cup \cdots \cup kH$$

is a decomposition of G into left cosets of H, prove that

$$G = H \cup Ha^{-1} \cup Hb^{-1} \cup \cdots \cup Hk^{-1}$$

is a decomposition into right cosets.

4. Decompose the order-8 group D_4 into left and right cosets of each of its nontrivial proper subgroups. Point out the subgroups for which left-coset decomposition is the same as right-coset decomposition. Compare those with the invariant subgroups of D_4. (See Problem 5 of Section 3.1.)

5. The same as Problem 4 but for the order-8 group Q_4. (See Problem 6 of Section 3.1.)

6. Decompose the group of nonzero real numbers under multiplication into cosets of each of the following subgroups: (a) all positive numbers; and (b) $\{1, -1\}$.

7. Decompose the group of nonsingular complex $n \times n$ matrices under matrix multiplication into cosets of each of the following invariant subgroups: (a) the group of unimodular complex $n \times n$ matrices: and (b) the kernel of its homomorphism to the group of unimodular complex $n \times n$ matrices. (See the second example and Problem 7 of Section 3.1.)

3.3. Factor Group

In the preceding section we found that decomposition of a group into left cosets of an invariant subgroup is the same as decomposition into right cosets of the same invariant subgroup. That fact has far-reaching consequences, as we will see in the present section and the next.

We now define a composition of cosets of invariant subgroup H of group G. It is denoted $(aH)(bH)$ and is defined as the set of all elements $h'h''$, where h' runs over all elements of coset aH and h'' independently runs over all elements of coset bH. That can also be expressed as the set of elements $(ah_1)(bh_2) = ah_1bh_2$, where h_1 and h_2 independently run over all elements of H. Using associativity and equality of left and right cosets of H, we obtain the following string of equations for the composition of cosets aH and bH, as defined just now:

$$(aH)(bH) = a(Hb)H$$

$$= a(bH)H$$

$$= (ab)(HH)$$

$$= (ab)H.$$

The last equality follows from $HH = H$, since HH is the set of elements $h_1 h_2$, where h_1 and h_2 independently run over all elements of H, which is just H itself, H being a subgroup. An expression such as $a(Hb)H$ is the set of elements $ah'h$, where h' runs over all elements of Hb and h independently runs over all elements of H, or equivalently, the set of elements $a(h_1 b)h_2 = ah_1 bh_2$, where h_1 and h_2 independently run over all elements of H. After overcoming our unfamiliarity with the notation, the point of what we are doing here can be appreciated; the composition of coset aH and coset bH, $(aH)(bH)$, is also a coset of H, the coset $(ab)H$ (although some element other than ab might be used to form it in the decomposition of G). Thus the set of cosets of any invariant subgroup H of group G is closed under coset composition.

That being the case, let us see if coset composition has any additional interesting properties. Is it associative? Yes,

$$[(aH)(bH)](cH) = (aH)[(bH)(cH)]$$

as a direct result of associativity of the composition of G. The composition of any coset of H with H itself gives

$$H(aH) = (Ha)H$$

$$= (aH)H$$

$$= a(HH)$$

$$= aH$$

and

$$(aH)H = a(HH) = aH,$$

using associativity, equality of left and right cosets, and $HH = H$. So H has the characteristic property of an identity in coset composition; its composition with any coset in either order is just that coset. And what happens when we compose any coset aH with coset $a^{-1}H$?

$$(aH)(a^{-1}H) = a(Ha^{-1})H$$

$$= a(a^{-1}H)H$$

$$= (aa^{-1})(HH)$$

$$= eH$$

$$= H$$

and

$$(a^{-1}H)(aH) = a^{-1}(Ha)H$$
$$= a^{-1}(aH)H$$
$$= (a^{-1}a)(HH)$$
$$= eH$$
$$= H,$$

using associativity, equality of left and right cosets, properties of inverse and identity, and $HH = H$. Thus for any coset aH there exists a coset $a^{-1}H$ (although some element other than a^{-1} might be used to form it in the decomposition of G), such that their composition in either order is H. In other words, every coset has an inverse coset.

What do we have, if not a new group? We start with any group G and invariant proper subgroup H. We decompose G into cosets of H (left and right cosets being the same),

$$G = H \cup aH \cup bH \cup \cdots .$$

We consider the set consisting of H and its cosets, $\{H, aH, bH, \ldots\}$ (Is it clear that each element of this set is itself a set, i.e., this is a set of sets?), under coset composition. And, indeed, we found that we have (1) closure, (2) associativity, (3) existence of identity, (4) existence of inverses—in short, a group. That group is called the *factor group* (also *quotient group*) of G by H and is denoted G/H. (It is a good idea to review the development of the factor group and make sure it is clear how H's being a subgroup and the invariance of H each plays its role.) If G is of finite order n and H is of order m, the order of the factor group G/H is $s = n/m$ by Lagrange's theorem.

As an example we take for G the non-Abelian, order-6 group D_3 and its invariant subgroup $H = \{e, a, b\}$. The corresponding coset decomposition is

$$G = \{e, a, b\} \cup \{c, d, f\}.$$

The factor group G/H is the set of sets

$$G/H = \{\{e, a, b\}, \{c, d, f\}\}$$

under coset composition. It is an order-2 group, isomorphic with C_2.

PROBLEM

Construct and identify (by finding an isomorphism with a more familiar group) the factor group G/H, where G and H are, respectively: (a) D_4 and each of its invariant subgroups; (b) Q_4 and each of its invariant subgroups; (c) the nonzero real numbers under multiplication and the positive real numbers; (d) the nonzero real numbers under multiplication and $\{1, -1\}$; (e) the nonsingular complex $n \times n$ matrices under matrix multiplication and the unimodular complex $n \times n$ matrices; and (f) the nonsingular complex $n \times n$ matrices under matrix multiplication and the kernel of the homomorphism of that group with the group of unimodular complex $n \times n$ matrices. (See Problems 4–7 of the preceding section.)

3.4. Anatomy of Homomorphism

Let us return once more to homomorphism $G \to G'$ with kernel K. As was proved in Section 3.1, K is an invariant proper subgroup of G (excluding the trivial homomorphism $G \to C_1$). So we decompose G into cosets of K,

$$G = K \cup aK \cup bK \cup \cdots ,$$

and construct the factor group

$$G/K = \{K, aK, bK, \ldots \}.$$

All elements of K are, by definition, mapped by the homomorphism to e', the identity element of G'. Consider the following structure-preservation diagram for any element a of G and all elements k of K, where a' is the image of a in G':

$$
\begin{array}{ccc}
a & k & = (ak) \\
\downarrow & \downarrow & \downarrow \\
a' & e' = & a' \; .
\end{array}
$$

Thus all members of coset aK are mapped to a', and all members of coset bK are mapped to b', the image of b in G', and so on.

Can members of different cosets of K be mapped to the same element of G'? No. To see that assume that b is not a member of coset aK, which is a necessary and sufficient condition for cosets aK and bK to be different. We then want to prove that, if $a \to a'$ and $b \to b'$, we must have $a' \neq b'$. So we assume the opposite, that $b \to a'$. By preservation of structure the image of b^{-1} in G' must be a'^{-1}. That gives us the structure-preservation diagram for the composition $b^{-1}a$:

$$
\begin{array}{ccc}
b^{-1} & a & = (b^{-1}a) \\
\downarrow & \downarrow & \downarrow \\
a'^{-1} & a' = & e' \; .
\end{array}
$$

Thus $b^{-1}a$ is an element of K, say $b^{-1}a = k$. Composition with b on the left and k^{-1} on the right gives $ak^{-1} = b$. But if k is an element of K, so is k^{-1} (since K is a subgroup), so that b has the form ak_1, where k_1 is an element of K, and b is a member of coset aK in contradiction to our assumption. Thus whereas all members of the same coset of K are mapped by the homomorphism to the same element of G', members of different cosets have different images in G'. And since, by definition, all members of G' are images in the homomorphism, we thereby obtain a one-to-one correspondence between all cosets of K in G and all elements of G'. We also obtain the result that any homomorphism is an m-to-one mapping, where m is the order of the kernel.

That one-to-one mapping between the factor group G/K and G' is, in fact,

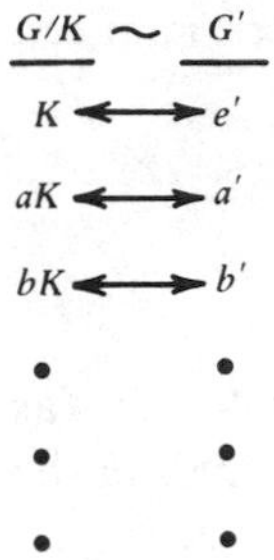

$$\frac{G/K \sim G'}{K \longleftrightarrow e'}$$

Fig. 3.1. Isomorphism of G/K with G', where K is kernel of homomorphism of G to G'.

$aK \longleftrightarrow a'$

$bK \longleftrightarrow b'$

$\bullet \qquad \bullet$

$\bullet \qquad \bullet$

$\bullet \qquad \bullet$

an isomorphism. Preservation of structure is based on the relation

$$(aK)(bK) = (ab)K,$$

found in the preceding section, and is exhibited in the diagram

$$(aK)(bK) = (ab)K$$
$$\downarrow \quad \downarrow \qquad \downarrow$$
$$a' \quad b' \ = (a'b')$$

where a' and b' are the respective images of a and b in G'. The mapping looks like that of Fig. 3.1.

To summarize the result, given a homomorphism $G \to G'$ with kernel K, we have the isomorphism $G/K \sim G'$.

A converse result is obtained, if, instead of starting with a homomorphism, we start with group G and invariant proper subgroup H. Decompose G into cosets of H,

$$G = H \cup aH \cup bH \cup \cdots,$$

and construct the factor group

$$G/H = \{H, aH, bH, \ldots\}.$$

Consider the mapping from G to G/H, where each element of G is mapped to that coset of H of which it is a member. That is an m-to-one mapping, where m is the order of H, in which every element of G has an image in G/H and every element of G/H serves as an image. That mapping is a homomorphism. To see that structure is preserved note that the cosets of H to which elements a, b, ab of G belong can be represented as $aH, bH, (ab)H$, respectively (although different elements might be used to form them). Then, by the relation

$$(aH)(bH) = (ab)H,$$

proved in the preceding section, we have the structure-preservation diagram

$$a \quad b \ = \ (ab)$$
$$\downarrow \quad \downarrow \qquad \downarrow$$
$$(aH)(bH) = (ab)H.$$

The kernel of that homomorphism is H, since all elements of H, and only those, are mapped to H, the identity element of G/H.

To sum up this section, given a homomorphism $G \to G'$ with kernel K, we have the isomorphism $G/K \sim G'$. Conversely, given group G and proper invariant subgroup H, we have the homomorphism $G \to G/H$ with kernel H.

That should help clarify what goes on in a homomorphism and how homomorphism, kernel, invariant proper subgroup, and factor group are intimately interrelated.

Some examples follow to help clarify the clarification.

1. We again return to the non-Abelian, order-6 group D_3, discussed so often above, as G. Its only invariant proper subgroup is $H = \{e, a, b\}$, C_3. The factor group G/H, here D_3/C_3, is the order-2 group

$$G/H = \{H, cH\}$$

$$= \{\{e, a, b\}, \{c, d, f\}\},$$

isomorphic with the abstract order-2 group $G' = \{e', a'\}$, C_2. We then have the three-to-one homomorphism $G \to G/H$ with kernel H, as well, of course, as the three-to-one homomorphism $G \to G'$ with kernel H. The whole situation can be represented as in Fig. 3.2.

2. Consider the order-4 group C_4, or its realization by the group of rotations about a common axis by $\{0°, 90°, 180°, 270°\}$, as G. Since it is Abelian, its only proper subgroup $H = \{e, b\}$, C_2, or its realization by the subgroup of rotations by $\{0°, 180°\}$, is invariant. The factor group, C_4/C_2, is

$$G/H = \{H, aH\} = \{\{e, b\}, \{a, c\}\},$$

or its rotational realization

$$\{\{0°, 180°\}, \{90°, 270°\}\}.$$

It is isomorphic with the abstract group C_2, $G' = \{e', a'\}$, or its realization by rotations by $\{0°, 180°\}$. We then have the two-to-one homomorphism $G \to G/H$ with kernel H and the two-to-one homomorphism $G \to G'$ with kernel H, among the abstract groups as well as among their rotational realizations. All that can be expressed by the diagram of Fig. 3.3

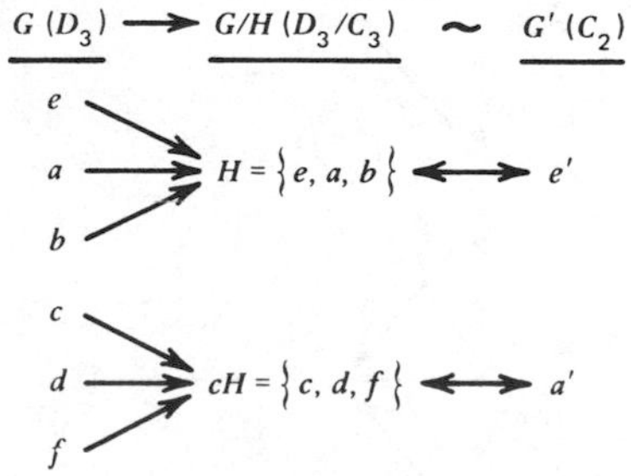

Fig. 3.2. Homomorphism of G (D_3) to G/H (D_3/C_3), isomorphism of G/H (D_3/C_3) with G' (C_2), and homomorphism of G (D_3) to G' (C_2).

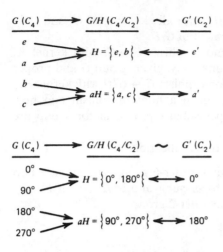

Fig. 3.3. Homomorphism of G (C_4) to G/H (C_4/C_2), isomorphism of G/H (C_4/C_2) with G' (C_2), and homomorphism of G (C_4) to G' (C_2): abstract groups and rotational realizations.

3. We take an infinite-order example. The group of all nonzero real numbers under multiplication, R, includes the group of all positive numbers under multiplication, P, as a subgroup. P is an invariant subgroup, since R is Abelian. The corresponding coset decomposition of R is $R = P \cup a \times P$, where a is any number not in P, i.e., any negative number. The coset $a \times P$ is just the set of all negative numbers, N. The order-2 factor group is $R/P = \{P, N\}$. It is isomorphic with any other order-2 group, C_2, say $Z = \{1, -1\}$ under multiplication. We then have the infinity-to-one homomorphisms $R \to R/P$ and $R \to Z$ with P as the kernel. Refer to Fig. 3.4.

PROBLEM

In light of the discussion of this section analyze the relations among G, H, and G/H for each part of the problem of the preceding section except part (c), which was analyzed in the third example of the present section.

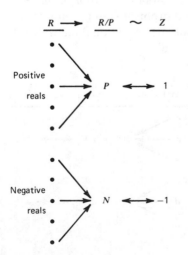

Fig. 3.4. Homomorphism of multiplicative group of all nonzero real numbers, R, to R/P, where P is multiplicative group of all positive numbers; isomorphism of R/P with multiplicative group $Z = \{1, -1\}$; and homomorphism of R to Z. N is set of all negative numbers.

3.5. Generators

Any subset of elements of a finite-order group such that they, their positive powers, and their compositions (and repeated compositions, as many times as necessary) produce the whole group is called a *set of generators* of the group. For a given group a set of generators is not unique. For example, the set of all elements of a group is a set of generators, and so is the set of all elements excluding the identity. However, there are sometimes smaller sets, and there are always minimal sets. Minimal sets of generators are not in general unique either. If a minimal set of generators of a group is a single element, so that this element and its positive powers produce the whole group, such a group is called a *cyclic group*, and this element is called a *generating element* of the group. Since all elements of a cyclic group are powers of a single element, a cyclic group is always Abelian. It can be shown that a group of prime order must be cyclic. Thus there is only a single abstract group of any prime order, the cyclic group of that order, which is Abelian. Finite-order groups of rotations about a common axis are cyclic. A possible generating element is the smallest (nonzero) rotation of each group. Cyclic groups are denoted C_n, where n is the order. Thus C_n can be realized by the group of rotations about a common axis by $\{0°, 360°/n, 2 \times 360°/n, \ldots, (n-1) \times 360°/n\}$. (The notation D_n is for the *dihedral groups*, of order $2n$. We do not discuss those here but present their general definition in Chapter 4.)

Examples of minimal sets of generators and cyclic groups include the following.

1. The order-2 group C_2 is cyclic, since $a^2 = e$, so a is its generating element. For its realization by $\{1, -1\}$ under multiplication the number -1 is the generating element, since $(-1) \times (-1) = 1$. In terms of its realization by rotations about a common axis by $\{0°, 180°\}$, rotation by $180°$ is the generating element, since two consecutive rotations by $180°$ produce a rotation by $360°$, the identity transformation.

2. The order-3 group C_3 is cyclic, since $a^2 = b$ and $a^3 = e$, or $b^2 = a$ and $b^3 = e$, so that a and b are each a generating element. In terms of its realization by rotations by $\{0°, 120°, 240°\}$ about a common axis, rotation by $120°$ is a generating element, since two consecutive rotations by $120°$ produce a rotation by $240°$, and three consecutive rotations result in a rotation by $360°$, which is rotation by $0°$. Rotation by $240°$ is also a generating element. Two consecutive rotations by $240°$ produce rotation by $480°$, which is rotation by $120°$, and three conseuctive $240°$ rotations make a rotation by $720° = 2 \times 360°$, which is the identity. Note that both 2 and 3 are primes and that there is only a single group of each order and it is cyclic.

3. The order-4 group C_4 is cyclic, since $a^2 = b$, $a^3 = c$, and $a^4 = e$, or $c^2 = b$, $c^3 = a$, and $c^4 = e$. Thus a and c are generating elements. However, b is not a generating element, since $b^2 = e$ and higher powers of b will not produce a and c. In terms of that group's realization by rotations about a

common axis by $\{0°, 90°, 180°, 270°\}$, the generating elements are rotation by 90° and rotation by 270°. Rotation by 180° is not a generating element, since two consecutive rotations by 180° produce the identity, and the other rotations are not produced by consecutive rotations by 180°.

4. Since 4 is not a prime, there may be noncyclic groups of that order, and indeed there is one, D_2. Minimal sets of generators of D_2 are $\{a, b\}$ ($a^2 = e$, $ab = c$), $\{a, c\}$ ($a^2 = e$, $ac = b$), and $\{b, c\}$ ($b^2 = e$, $bc = a$). In terms of its realization by reflections and rotation, presented in Section 2.1, a minimal set of generators consists of two reflections or the rotation and either reflection.

5. The order-5 group C_5 has prime order and is cyclic. Any one of its nonidentity elements is a generating element, for example, c, $c^2 = a$, $c^3 = d$, $c^4 = b$, $c^5 = e$. In the rotational realization of that group any nonzero rotation is a generating element.

6. The order-6 group D_3 is non-Abelian and therefore not cyclic. Two of its minimal sets of generators are $\{a, c\}$ ($a^2 = b$, $a^3 = e$, $ac = f$, $a^2c = d$) and $\{d, f\}$ ($d^2 = e$, $df = a$, $fd = b$, $dfd = c$).

PROBLEMS

1. Prove that a group of prime order must be cyclic.

2. Close your eyes and find all abstract order-7 groups. Display their group tables.

3. Construct the group table of the cyclic group of order 8, C_8.

4. Find at least two minimal sets of generators for the order-8 group D_4 and construct all the group elements from each set. (See Problem 6 of Section 2.6.)

5. $\{a, b\}$ is a minimal set of generators of a certain group, where $a^4 = b^2 = e$ and $ba = a^3b$. Construct the composition table for the set $\{e, a, a^2, a^3, b, ab, a^2b, a^3b\}$ and show that the set is closed. Since the set contains $\{a, b\}$ and is closed, it is the group under consideration. Show that this is D_4. (See Problem 6 of Section 2.6.)

6. The same as Problem 5, except that $a^4 = e$, $a^2 = b^2$, $ba = a^3b$, and the group is Q_4. (See Problem 7 of Section 2.6.)

3.6. Direct Product

The *direct product* of two groups G and G' is denoted $G \times G'$ and is the group consisting of elements that are ordered pairs (g, g'), where g and g' run independently over all elements of G and G', respectively. If the orders of G and G' are n and n', respectively, the order of $G \times G'$ is nn'. Composition in $G \times G'$ is defined

$$(a, a')(b, b') = (ab, a'b')$$

for all a, b in G and all a', b' in G'. That assures (1) closure and (2) associativity. (3) The identity of $G \times G'$ is (e, e'), where e and e' are the respective identities of G and G'. (4) The inverse of any element (g, g') of

(e, e)	(e, a)	(a, e)	(a, a)
(e, a)	(e, e)	(a, a)	(a, e)
(a, e)	(a, a)	(e, e)	(e, a)
(a, a)	(a, e)	(e, a)	(e, e)

Fig. 3.5. Group table of $C_2 \times C_2$.

$G \times G'$ is (g^{-1}, g'^{-1}). Thus $G \times G'$ is proved to be a group as claimed. Among the possible proper subgroups of $G \times G'$ there is one isomorphic with G, consisting of elements of the form (g, e') for all g in G, and another isomorphic with G', consisting of elements of the form (e, g') for all g' in G'. Both are invariant. Clearly the direct-product group $G' \times G$ is isomorphic with $G \times G'$. Higher-degree direct-product groups are formed similarly. For instance, $G \times G' \times G''$ is the group of ordered triples (g, g', g''), where g, g', g'' run independently over all elements of G, G', G'', respectively. Composition is

$$(a, a', a'')(b, b', b'') = (ab, a'b', a''b'').$$

Consider the following examples.

1. The direct product $C_2 \times C_2$ is an order-4 group. But which? The elements of that group are $(e, e), (e, a), (a, e)$, and (a, a), with $a^2 = e$. Figure 3.5 is the group table. The total domination of the group-table diagonal by the identity element discloses that this is the structure of D_2. One possible isomorphism mapping is

$$(e, e), (e, a), (a, e), (a, a)$$
$$\updownarrow \quad \updownarrow \quad \updownarrow \quad \updownarrow$$
$$e', \quad a', \quad b', \quad c' \,.$$

2. The direct product $C_2 \times C_3$ is an order-6 group. Since the constituent groups are Abelian, so is their direct product. Therefore, that is not the order-6 group D_3, discussed so frequently already. Its elements are (e, e'), $(a, a'), (e, b'), (a, e'), (e, a'), (a, b')$, where $\{e, a\}$ forms the order-2 group and $\{e', a', b'\}$ makes up the order-3 group. The group table is seen in Fig. 3.6. If we relabel the elements $e'', a'', b'', c'', d'', f''$, respectively, the group table

(e, e')	(a, a')	(e, b')	(a, e')	(e, a')	(a, b')
(a, a')	(e, b')	(a, e')	(e, a')	(a, b')	(e, e')
(e, b')	(a, e')	(e, a')	(a, b')	(e, e')	(a, a')
(a, e')	(e, a')	(a, b')	(e, e')	(a, a')	(e, b')
(e, a')	(a, b')	(e, e')	(a, a')	(e, b')	(a, e')
(a, b')	(e, e')	(a, a')	(e, b')	(a, e')	(e, a')

Fig. 3.6. Group table of $C_2 \times C_3$.

e''	a''	b''	c''	d''	f''
a''	b''	c''	d''	f''	e''
b''	c''	d''	f''	e''	a''
c''	d''	f''	e''	a''	b''
d''	f''	e''	a''	b''	c''
f''	e''	a''	b''	c''	d''

Fig. 3.7. Group table of C_6.

takes the form of Fig. 3.7. That is clearly the cyclic group of order 6, C_6. Its generating elements are a'' and f'', or in $C_2 \times C_3$ notation (a, a') and (a, b'). C_6 can be realized by the group of rotations about a common axis by $\{0°, 60°, 120°, 180°, 240°, 300°\}$, whose generating elements are rotation by $60°$ and rotation by $300°$. C_6 and D_3 are the only abstract groups of order 6.

3. The group of complex numbers under addition is isomorphic with the direct product of the group of real numbers under addition with itself.

4. The group of 3-vectors under vector addition is isomorphic with the direct product of the group of real numbers under addition with itself twice.

PROBLEMS

1. In the direct-product group $G \times G'$ prove that the two subgroups consisting, respectively, of elements of the form (g, e') and of elements of the form (e, g'), isomorphic with G and G', respectively, are both invariant. Decompose $G \times G'$ into cosets of each. Identify the factor groups $G \times G'/G$ and $G \times G'/G'$.

2. Construct the group table of the order-8 group $C_2 \times C_4$. Find a minimal set of generators and reconstruct the group table in terms of it. Prove that this group is not isomorphic with C_8, D_4, or Q_4.

3. Construct the group table of the order-8 group $C_2 \times C_2 \times C_2$. Find a minimal set of generators and reconstruct the group table in terms of it. Prove that this group is not isomorphic with C_8, $C_2 \times C_4$, D_4, or Q_4.

4. We have now collected five order-8 groups: C_8, $C_2 \times C_4$, $C_2 \times C_2 \times C_2$, D_4, and Q_4. Those are, in fact, the only abstract groups of order 8. So the order-8 group $C_2 \times D_2$ must be isomorphic with one of them. With which?

3.7. Permutations, Symmetric Groups

A *permutation* is a rearrangement of objects. Imagine, for example, that we have four objects in positions 1, 2, 3, 4, and imagine that we permute (i.e., rearrange) them so that the object in position 1 is placed in position 4, the object in position 2 is placed in position 1, the object in position 3 is left where it is, and the object in position 4 is placed in position 2. Figure 3.8

Fig. 3.8. Permutation $\begin{pmatrix} 1234 \\ 4132 \end{pmatrix}$.

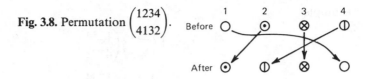

should help. Such a permutation is denoted $\begin{pmatrix} 1234 \\ 4132 \end{pmatrix}$, where the top row lists the positions, and each number in the bottom row is the final location of the object originally located at the position whose number is directly above it in the top row. The top row can be in any order, so both $\begin{pmatrix} 2431 \\ 1234 \end{pmatrix}$ and $\begin{pmatrix} 3142 \\ 3421 \end{pmatrix}$, for instance, also denote the permutation in our example. A general permutation of n objects can be denoted

$$\begin{pmatrix} 1 & 2 & \cdots n \\ p_1 p_2 & \cdots p_n \end{pmatrix},$$

where $p_1, p_2, \ldots, p_n$ are the numbers $1, 2, \ldots, n$ is some order (i.e., some permutation of $1, 2, \ldots, n$), which directs us to move the object in position 1 to position p_1, the object in position 2 to position p_2, and so on. The top row can be in any order, so long as p_i appears under i for $i = 1, 2, \ldots, n$.

Permutations of the same number of objects are composed by consecutive application; the second permutations acts on the result of the first. The composition of permutations $a = \begin{pmatrix} 1234 \\ 4132 \end{pmatrix}$ and $b = \begin{pmatrix} 1234 \\ 1342 \end{pmatrix}$, for example, with permutation a acting first and permutation b acting on the result of a, is denoted

$$ba = \begin{pmatrix} 1234 \\ 1342 \end{pmatrix}\begin{pmatrix} 1234 \\ 4132 \end{pmatrix}.$$

(The order of application of permutations, as well as of any other kind of transformation, is specified from right to left. That is discussed in Chapter 4.) The composition is the permutation obtained by following the itinerary of each object through the permutations to which it is subjected. In the example the object orginally in position 1 is carried to position 4 by permutation a and from there to position 2 by b. The object originally in position 2 is shifted to position 1 by a and then left in position 1 by b. And so on to obtain $ba = \begin{pmatrix} 1234 \\ 2143 \end{pmatrix}$. A useful trick is to reorder the rows of the second permutation to make the top row match the bottom row of the first permutation. The combined permutation is then obtained by "canceling" the bottom row of the first permutation with the top row of the second, leaving the top row of the first permutation and the bottom row of the second as the result. In the example

$$ba = \begin{pmatrix} 1234 \\ 1342 \end{pmatrix}\begin{pmatrix} 1234 \\ 4132 \end{pmatrix}$$

$$= \begin{pmatrix} 4132 \\ 2143 \end{pmatrix}\begin{pmatrix} 1234 \\ 4132 \end{pmatrix}$$

$$= \begin{pmatrix} 1234 \\ 2143 \end{pmatrix}.$$

Also

$$ab = \begin{pmatrix} 1234 \\ 4132 \end{pmatrix}\begin{pmatrix} 1234 \\ 1342 \end{pmatrix}$$

$$= \begin{pmatrix} 1342 \\ 4321 \end{pmatrix}\begin{pmatrix} 1234 \\ 1342 \end{pmatrix}$$

$$= \begin{pmatrix} 1234 \\ 4321 \end{pmatrix}.$$

For general permutations of n objects

$$a = \begin{pmatrix} 1 & 2 & \cdots n \\ p_1 p_2 & \cdots p_n \end{pmatrix} \quad \text{and} \quad b = \begin{pmatrix} 1 & 2 & \cdots n \\ q_1 q_2 & \cdots q_n \end{pmatrix}$$

the composition is

$$ba = \begin{pmatrix} 1 & 2 & \cdots n \\ q_1 q_2 & \cdots q_n \end{pmatrix}\begin{pmatrix} 1 & 2 & \cdots n \\ p_1 p_2 & \cdots p_n \end{pmatrix}$$

$$= \begin{pmatrix} p_1 & p_2 & \cdots p_n \\ q_{p_1} q_{p_2} & \cdots q_{p_n} \end{pmatrix}\begin{pmatrix} 1 & 2 & \cdots n \\ p_1 p_2 & \cdots p_n \end{pmatrix}$$

$$= \begin{pmatrix} 1 & 2 & \cdots n \\ q_{p_1} q_{p_2} & \cdots q_{p_n} \end{pmatrix}.$$

Similarly,

$$ab = \begin{pmatrix} 1 & 2 & \cdots n \\ p_{q_1} p_{q_2} & \cdots p_{q_n} \end{pmatrix}.$$

Thus the composition of permutations of n objects is also a permutation of n objects. Composition of permutations is not, in general, commutative, as we see in the example or in the general case. It is, however, associative, as we could easily prove by composing three permutations and obtaining three tiers of indices in the bottom row, which we mercifully refrain from doing. (The composition of transformations of any kind by consecutive application is associative, as proved in Chapter 4.)

The "nonpermutation" $\begin{pmatrix} 12 \cdots n \\ 12 \cdots n \end{pmatrix}$ serves as the identity permutation of n

$$S_3 = \left\{ \begin{pmatrix} 123 \\ 123 \end{pmatrix}, \begin{pmatrix} 123 \\ 231 \end{pmatrix}, \begin{pmatrix} 123 \\ 312 \end{pmatrix}, \begin{pmatrix} 123 \\ 132 \end{pmatrix}, \begin{pmatrix} 123 \\ 321 \end{pmatrix}, \begin{pmatrix} 123 \\ 213 \end{pmatrix} \right\}$$

$$\wr \qquad \updownarrow \qquad \updownarrow \qquad \updownarrow \qquad \updownarrow \qquad \updownarrow \qquad \updownarrow$$

$$D_3 = \{ \quad e, \qquad a, \qquad b, \qquad c, \qquad d, \qquad f, \ \}$$

Fig. 3.9. Isomorphism of S_3 with D_3.

objects. The inverse of permutation

$$\begin{pmatrix} 1 & 2 & \cdots n \\ p_1 p_2 & \cdots p_n \end{pmatrix}$$

is simply

$$\begin{pmatrix} p_1 p_2 & \cdots p_n \\ 1 & 2 & \cdots n \end{pmatrix};$$

each clearly undoes the other.

What could we be leading up to, if not that the set of all permutations of n objects, under composition of consecutive permutation, forms a group? It is called the *symmetric group* of degree n and is denoted S_n. In the preceding paragraphs we (1) proved closure, (2) claimed associativity, and proved (3) existence of identity and (4) existence of inverses. The order of S_n in $n!$. It is non-Abelian for $n > 2$.

It should be clear that $S_n \supset S_m$ for $n \geq m$.

Examples of symmetric groups include the following.

1. S_1 is the order-1 group of the identity permutation $\begin{pmatrix} 1 \\ 1 \end{pmatrix}$.

2. S_2 consists of the permutations $\begin{pmatrix} 12 \\ 12 \end{pmatrix}$ and $\begin{pmatrix} 12 \\ 21 \end{pmatrix}$ and is isomorphic with C_2. It is Abelian.

3. S_3 is an order-6 group. It is non-Abelian, so it must be isomorphic with D_3. (See the second example of the preceding section.) A mapping is given in Fig. 3.9. (See Fig. 2.9.)

4. S_4 is of order 24, too large to consider here, and higher-degree symmetric groups have even larger, and rapidly increasing, orders. For example, $11! = 39,916,800$, and $12!$ exceeds the capacity of an eight-digit calculator.

PROBLEMS

1. Prove what was declared in the text as being clear, that $S_n \supset S_m$ for $n \geq m$.

2. Given the permutations $a = \begin{pmatrix} 12345 \\ 53214 \end{pmatrix}$ and $b = \begin{pmatrix} 12345 \\ 32154 \end{pmatrix}$: (a) Calculate the permutations a^2, b^2, ab, and ba; (b) find the permutations a^{-1}, b^{-1}, $(ab)^{-1}$, and $(ba)^{-1}$, calculate $b^{-1}a^{-1}$ and $a^{-1}b^{-1}$, and compare; (c) solve for permutations c and d, where $ac = b = da$; and (d) solve for permutations f and g, where $bf = a = gb$.

3. Check associativity of composition of permutations by comparing $(ab)c$ and $a(bc)$ for the special case of $a = \begin{pmatrix} 123456 \\ 415263 \end{pmatrix}$, $b = \begin{pmatrix} 123456 \\ 361542 \end{pmatrix}$, $c = \begin{pmatrix} 123456 \\ 146352 \end{pmatrix}$.

4. Calculate all compositions of elements of S_3 and construct a group table for that group.

3.8. Cycles, Transpositions, Alternating Groups

For another way of looking at permutations we consider the permutation

$$\begin{pmatrix} 12345678 \\ 45132768 \end{pmatrix},$$

an element of S_8. The object originally in position 1 is moved to position 4. The object originally in position 4 must be evicted and is placed in position 3. The object evicted from position 3 is moved into position 1, closing the circle. Such a chain of replacements is called a *cycle*. That cycle is $1 \to 4 \to 3 \to 1$. It is denoted (143). The other cycles of our permutation are $2 \to 5 \to 2$, denoted (25), $6 \to 7 \to 6$, denoted (67), and $8 \to 8$, denoted (8). A cycle is a permutation, a cyclic permutation. For example,

$$(143) = \begin{pmatrix} 12345678 \\ 42135678 \end{pmatrix}, \qquad (25) = \begin{pmatrix} 12345678 \\ 15342678 \end{pmatrix}, \qquad (8) = \begin{pmatrix} 12345678 \\ 12345678 \end{pmatrix}.$$

Since a cycle is a closed replacement chain, it may be written starting with any location in the chain. For example, $(143) = (431) = (314)$ and $(25) = (52)$. However, the cyclic order must be kept. For example, $(143) \neq (134)$.

Thus any permutation consists of a number of cycles. Those cycles all commute with each other, since they apply to different locations, i.e., no two cycles of a permutation contain the same symbol. Any permutation can be denoted as a composition of its cycles, in any order since they commute. For our example

$$\begin{pmatrix} 12345678 \\ 45132768 \end{pmatrix} = (67)(143)(25)(8)$$

or just $(67)(143)(25)$, since a cycle of one is just the identity and can be dropped (as long as the degree of the permutation is remembered). Or, equally well,

$$\begin{pmatrix} 12345678 \\ 45132768 \end{pmatrix} = (431)(67)(52)$$

$$= (25)(314)(76), \quad \text{etc.}$$

The decomposition of a permutation into cycles clearly is unique (although it may be denoted in more than one way).

A cycle of two, an interchange of two objects, is called a *transposition*. Any cycle can be decomposed into transpositions. For example,

$$(143) = (13)(14) = (41)(43).$$

(Read transposition compositions from right to left!) In such decompositions
the transpositions do not commute, since they have symbols in common, so
their order is important. Also, as seen in the example, decomposition into
transpositions is not unique. But an n-cycle always decomposes into a minimum of $n - 1$ transpositions. Here is a recipe for such a decomposition:

$$(12\cdots n) = (1n)\cdots(13)(12).$$

And if an n-cycle is decomposed into more than $n - 1$ transpositions, it turns
out that the number of transpositions can differ from $n - 1$ only by a multi-
ple of 2. Thus an n-cycle can be decomposed only into an odd number of
transpositions if n is even, or an even number of transpositions if n is odd.

In that way any permutation can be decomposed into a composition of
transpositions; it is first decomposed into cycles, then all the cycles are decomposed into transpositions. For example,

$$\begin{pmatrix} 12345678 \\ 45132768 \end{pmatrix} = (67)(143)(25)$$

$$= (67)(13)(14)(25).$$

The decomposition is not unique, and neither is the number of transpositions
in the decomposition. What is uniquely determined, however, is the minimal
number of transpositions in the decomposition and the evenness or oddness
of the number of transpositions in the decomposition. If the number is even
or odd, the permutation is said to be, respectively, an even or odd permutation, or to possess, respectively, even or odd *parity*. [You might be familiar
with a definition of the determinant of a square matrix in which the sign of
each term is determined by the parity of the permutation of indices in that
term through the function $(-1)^h$, where h is the number of transpositions
involved in the permutation.]

The composition of two even or two odd permutations is an even permutation. The composition of an even and an odd permutation is an odd
permutation. The identity permutation is even (zero transpositions). The inverse of an even or odd permutation is, respectively, even or odd. Thus the
set of all even permutations of n objects forms a subgroup of S_n (the odd
permutations do not), called the *alternating group* of degree n and denoted
A_n. The order of A_n is $\frac{1}{2}n!$ A_n is, in fact, an invariant subgroup of S_n. The
factor group S_n/A_n is isomorphic with C_2.

For example, in Fig. 3.10 we decompose the elements of S_3 into transpositions and indicate their parities. For convenience we also show the
isomorphism mapping to D_3 (see Fig. 2.9). The invariant subgroup A_3 of S_3 is
isomorphic with and mapped to the invariant subgroup $\{e, a, b\}$, C_3, of D_3.

We use that example to elucidate the significance of conjugacy among

$D_3 \sim S_3$	Parity	
$e \leftrightarrow \begin{pmatrix} 123 \\ 123 \end{pmatrix} = (1)$	Even	
$a \leftrightarrow \begin{pmatrix} 123 \\ 231 \end{pmatrix} = (123) = (13)(12)$	Even	$\left. \right\} A_3 \sim C_3$
$b \leftrightarrow \begin{pmatrix} 123 \\ 312 \end{pmatrix} = (132) = (12)(13)$	Even	
$c \leftrightarrow \begin{pmatrix} 123 \\ 132 \end{pmatrix} = (23)$	Odd	
$d \leftrightarrow \begin{pmatrix} 123 \\ 321 \end{pmatrix} = (13)$	Odd	
$f \leftrightarrow \begin{pmatrix} 123 \\ 213 \end{pmatrix} = (12)$	Odd	

Fig. 3.10. Decomposition of elements of S_3 into transpositions with parities indicated. Isomorphism of S_3 with D_3 is shown for convenience. Set of even-parity elements, A_3, is invariant subgroup of S_3 and is isomorphic with C_3, the invariant subgroup of D_3.

elements of S_n. For D_3 we found the conjugacy classes e, $a \equiv b$, $c \equiv d \equiv f$. In terms of its realization by S_3, the conjugacy classes are (1), $(123) \equiv (132)$, $(23) \equiv (13) \equiv (12)$. Thus here and for general S_n, conjugate permutations have the same cycle structure. They differ only in the labeling of the symbols that are permuted. That is analogous with similarity transformations in linear algebra. There square, column, and row matrices related by a similarity transformation can be interpreted as representing the same object, an operator or a vector, but expressed with respect to different bases, i.e., different "names" or "labels" for the same objects.

PROBLEMS

1. Prove the statements made in the text: (a) The composition of two even or two odd permutations is an even permutation; (b) the composition of an even and an odd permutation is an odd permutation; (c) the parity of permutation a is the same as that of a^{-1}; and (d) A_n is an *invariant* subgroup of S_n.

2. Referring to Problem 2 of Section 3.7, take the results obtained there and: (a) express a, b, a^2, b^2, ab, ba in cycle form, then recalculate the compositions in cycle representation; (b) express a, b, a^2, b^2, ab, ba in the form of transpositions, then recalculate the compositions in that representation; and (c) check the parities of a, b, a^2, b^2, ab, ba. (*Hints:* Remember that the order of operation of consecutive permutations is from right to left. When composing transpositions, the result is conveniently obtained in cycle form and then converted to transpositions. Remember that representation of a permutation in transposition form is not unique.)

3.9. Cayley's Theorem

Consider the group table of any group G of finite order n. For example, the group D_3, whose table is repeated in Fig. 3.11. If the first row is $e, a, b, \ldots$, the row starting with g is obtained from the first row by composition with g on the left: $ge \, (= g), ga, gb, \ldots$. From that it follows directly that (1) each row contains all the elements of the first row, but in some other arrangement (unless $g = e$), and (2) rows starting with different elements are different. As for (1), if any row did not contain all the elements of the first row, we would have, say, $ga = gb$ for the element g at the left of that row and some pair of elements a, b. Composing both sides of that equation with g^{-1} on the left, we would then have $a = b$, which is a contradiction, since the first row of the group table contains n different symbols for n distinct elements. The arrangement of the elements in the row starting with g is different from that of the first row, simply because the first row starts with e and the other does not (unless $g = e$, and we are talking about the first row anyway). The second result is trivial, since rows starting with different elements are different simply in that they start with different elements.

What we are getting at, however, is that to each element of a finite group we can assign a unique permutation of n objects in a one-to-one correspondence; to every element g we assign the permutation whose top row is the top row of the group table and whose bottom row is that row of the group table starting with g,

$$g \leftrightarrow \begin{pmatrix} e & a & b & \cdots \\ g & ga & gb & \cdots \end{pmatrix}.$$

(That is indeed a permutation, although we have become used to denoting permutations with numbers rather than with letters. To make it look more familiar one might replace the letters $e, a, b, \ldots$ by the numbers $1, 2, \ldots, n$ according to some cipher.) From the discussion of the preceding paragraph it is clear that this is indeed a one-to-one mapping of all elements of G onto a subset of S_n. In fact, that mapping is an isomorphism, and that subset of S_n is therefore a subgroup of S_n. To see that structure is preserved consider Fig. 3.12, keeping in mind our trick for composing permutations. (The whole discussion could be carried out just as well with columns instead of rows of a group table.) Our result, known as Cayley's theorem (Arthur Cayley, British

e	a	b	c	d	f
a	b	e	f	c	d
b	e	a	d	f	c
c	d	f	e	a	b
d	f	c	b	e	a
f	c	d	a	b	e

Fig. 3.11. Group table of D_3.

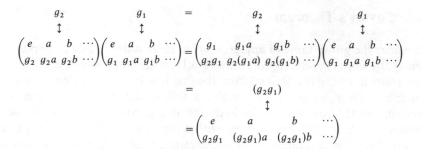

Fig. 3.12. Structure-preservation diagram for Cayley's theorem.

mathematician, 1821–1895), is that every group of finite order n is isomorphic with a subgroup of S_n. Cayley's theorem is helpful for finding groups of a given order.

Cayley's theorem also helps us elucidate the significance of conjugacy among group elements. We saw previously that conjugate permutations have the same cycle structure, i.e., they are, in a sense, the same permutation but differ only in the labeling of the permuted sybmols. By Cayley's theorem conjugate group elements are isomorphically mapped to conjugate permutations reflecting the action of those elements on the group itself through composition. Thus conjugate elements fulfill similar functions within the group. As an example, you might review our numerous analyses of the group D_3, comparing the functions of conjugate and nonconjugate elements in the various intragroup relations.

PROBLEM

(a) Display a realization of the order-6 group D_3 as a subgroup of the order-6! = 720 group S_6 by Cayley's theorem. (Note that all of S_3 realizes D_3 due to their isomorphism.) (b) Express the elements of that subgroup in cycle form and compare conjugate elements. (c) Express the elements in the form of transpositions and find their parity. Verify that all the even permutations realize the nontrivial invariant subgroup of D_3.

3.10. Summary of Chapter Three

In this chapter we continued the introduction to group theory and related mathematical topics that we started in the preceding chapter. We summarize by listing the more important concepts presented in each section.

In Section 3.1: conjugate elements, conjugacy class, equivalence class, invariant (normal) subgroup, kernel of homomorphism.

In Section 3.2: coset of group, coset decomposition, Lagrange's theorem.

In Section 3.3: coset composition, factor (quotient) group.

In Section 3.4: homomorphism.

In Section 3.5: set of generators, cyclic group, generating element.

In Section 3.6: direct product of groups.

In Section 3.7: permutation, symmetric group.

In Section 3.8: cycle, transposition, parity of permutation, alternating group.

In Section 3.9: Cayley's theorem.

CHAPTER FOUR

Symmetry: The Formalism

In Chapters 8–10 we study the general theory of symmetry in science via the conceptual approach. That approach is the best for understanding the concepts involved in symmetry and the significance of symmetry in science. But the *application* of symmetry considerations in science, and especially quantitative applications, require a general symmetry formalism. That formalism is developed in the present chapter and in Chapters 5–7 following. Mathematical concepts presented in the preceding two chapters are used extensively. Our road to symmetry starts with a state space of a system and proceeds through transformation, transformation group, equivalence relation for a state space, symmetry transformation, and symmetry group. The quantification of symmetry is discussed, and a special section is devoted to symmetry of quantum systems.

Except for the necessary mathematical concepts, presented in the preceding chapters, the general theory of symmetry is developed from scratch. For most of the examples and problems the mathematical background required for the examples and problems of the preceding chapters more than suffices, while some basic familiarity with physical phenomena is also desirable. However, for Section 4.7, Quantum Systems, and for some of the examples and problems in the other sections a good understanding of the Hilbert-space (David Hilbert, German mathematician, 1862–1943) formulation of quantum theory and of some rather advanced physics is a prerequisite. This material is definitely not for the "general reader," who is advised to skip it.

4.1. System, State

We start by presenting the concepts of system, subsystem, state, and state space. Those concepts are purposely not sharply defined and assumed to be intuitively understood, in order to allow the widest possible applicability of the following development.

A *system* is whatever we investigate the properties of. We are intentionally being vague and general here, so as to impose no limitations on the possible objects of our interest. A system might be abstract or concrete, microscopic or macroscopic, static or dynamic, finite or infinite. Anything (and not only things) can be a system!

A *subsystem* is a system wholly contained, in some sense, within a system. Again, vagueness and generality reign.

A *state* of a system is a possible condition of the system. Here too we are saying very little. We can add, though, that a system might have a finite or (denumerably or nondenumerably) infinite number of states, and that the same system might have different kinds or numbers of states, depending on how it is being considered.

A *state space* of a system is the set of all states of the same kind, where "the same kind" may be interpreted in any useful way.

Consider some examples.

1. Let the system be a given amount of a certain pure gas. Considered microscopically, the states of the system are characterized by the positions and momenta of all the molecules (assuming spinless, structureless, point particles). The set of all allowable positions and momenta for all molecules forms a state space of the system. Macrostates, assuming that the gas is confined and in equilibrium, are described by any two quantities among, say, pressure, volume, and temperature. So a state space might be the set of all pressures and temperatures. In both cases, microscopic and macroscopic, the number of states is infinite, though "more so" in the microscopic case.

2. Take as the system a plane figure. States of such a system are described by shape, size, location, orientation, and even color, texture, and so on, and are infinite in number.

3. Or to be more explicit, let the system be a plane figure of given shape and size, lying in a given plane, and with one of its points fixed in the plane. Ignore color, texture, and the like. States of this system are characterized by a single angle specifying its rotational orientation in the plane and by its *handedness* (also called *chirality*), i.e., which of its two mirror-twin versions (such as left and right hands) is appearing.

4. Consider the system of a ball lying in any one of three depressions in the sand. That system has three states.

5. Consider the system of three balls lying in three depressions. States are described by specifying which ball is in which depression. The state space of that system consists of six states, the six possible arrangements of three balls.

6. The system might be any quantum system. Its states are quantum states, characterized by a set of quantum numbers, the eigenvalues of a complete set of commuting Hermitian operators (after Charles Hermite, French mathematician, 1822–1901). State space is postulated to be a Hilbert space, in which states are represented by vectors (actually rays). Quantum systems are treated in more detail in Section 4.7.

PROBLEMS

1. Considering only properties such as number, color, and suit, describe various state spaces for a pair of playing cards drawn from the same deck.

2. You have a set of five different flags and three flagpoles, each of which may either be empty or fly one of the flags. How many states does this system have, if you are interested only in which flags are flying and do not care on which flagpoles? How many states if the positions of the flags are important?

3. The system is a fully unfurled flag flying from a flagpole. What are the system's states?

4. What are the states of a regular pentagon of given size, lying in a given plane, with its center fixed at a given point? (It might help to label *two* points.)

5. What are the states of a straight line lying in a given plane, say the xy-plane? Express your result analytically.

6. You have a mixture of the gases H_2, Cl_2, and HCl in a closed container of constant volume with given numbers of hydrogen and chlorine atoms. Describe the system's microstates. Assume that the mixture is in equilibrium, meaning dynamical equilibrium, since the reactions $H_2 + Cl_2 \rightleftarrows 2HCl$ are continually taking place. The relative rates of the combination and dissociation reactions are directly affected by temperature, pressure, and densities. Describe the macrostates of the system.

7. Describe the most inclusive state space for a system consisting of a free proton and a free neutron, where the particles have spin (i.e., a half unit of intrinsic angular momentum apiece) but are otherwise considered as being classical and non-relativistic point particles.

8. Describe the most inclusive state space for a system of two free nucleons (a nucleon is a proton or a neutron), where the total energy of the system equals some given value. The particles are considered as being point particles with spin, and the quantum rule for relative spin orientation is assumed to hold; otherwise the system is classical. Use relativistic kinematics.

9. What is the space of internal states for a hydrogen atom?

4.2. Transformations, Transformation Group

A *transformation* of a system is a mapping of a state space of the system into itself; i.e., every state has an image, which is also a state; more than one state might have the same image; but not every state must necessarily serve as an image in the mapping. We denote a transformation T by

$$u \xrightarrow{\ T\ } v \qquad \text{or} \qquad v = T(u)$$

(arrow notation or function notation), where u and v represent states of the system. State v is the image of state u under transformation T. State u runs over all states of a state space. If the number of states of a state space is finite, a transformation might be expressed as a two-column table. Otherwise it must be expressed as a general rule.

Transformations are composed by consecutive application, i.e., the composition of two transformations is defined as the result of applying one transformation to the result of the other. For transformations T_1 and T_2 one possible composition is $T_2 T_1$, denoting the result of first applying T_1 and then T_2:

$$u \xrightarrow{\ T_1\ } v \xrightarrow{\ T_2\ } w \qquad \text{or} \qquad w = T_2(v)$$

$$T_2 T_1$$

$$= T_2(T_1(u))$$

$$\overset{\text{def}}{=} (T_2 T_1)(u).$$

The function notation practically "forces" us to specify consecutively applied transformations from right to left, and we adhere to that convention. The other possible composition of T_1 and T_2 is, of course, $T_1 T_2$, where T_2 is applied first, followed by T_1. For composition of transformations to be meaningful, the transformations must act in the same state space. That will always be tacitly assumed. As defined, the composition of any two transformations is a mapping of a state space into itself and is therefore also a transformation. Thus the set of transformations of a system for any of its state spaces is closed under the composition of consecutive application.

In general

$$T_2 T_1 \neq T_1 T_2.$$

However, a pair of transformations T_1, T_2 such that

$$T_2 T_1 = T_1 T_2,$$

i.e.,

$$(T_2 T_1)(u) = (T_1 T_2)(u)$$

for all states u of a state space, are said to *commute*.

Composition of transformations by consecutive application is easily shown to be associative, i.e.,

$$T_3(T_2 T_1) = (T_3 T_2) T_1$$

for all transformations T_1, T_2, T_3, so that the notation $T_3 T_2 T_1$ is unambiguous. Using the definition of composition, the transformation on the left-hand side of this equation acting on arbitrary state u is

$$(T_3(T_2 T_1))(u) = T_3((T_2 T_1)(u))$$

$$= T_3(T_2(T_1(u))).$$

Similarly, the right-hand side gives

$$((T_3 T_2) T_1)(u) = (T_3 T_2)(T_1(u))$$

$$= T_3(T_2(T_1(u))),$$

which is the same. That proves associativity. In arrow notation, which is

more awkward here, the same proof is

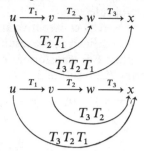

The mapping I of every state of a state space to itself,

$$u \xrightarrow{I} u \qquad \text{or} \qquad u = I(u)$$

for all states of a state space, is called the *identity transformation*. Indeed it acts as the identity under composition by consecutive application.

$$TI = IT = T$$

for all T, since for arbitrary state u

$$(TI)(u) = T(I(u)) = T(u),$$

$$(IT)(u) = I(T(u)) = T(u).$$

In arrow notation that is

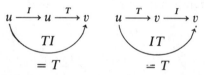

Among the transformations of a system there are those that are one-to-one and onto (bijective), i.e., every state of a state space serves as an image of the mapping and is the image of only a single state. For those transformations, and only for those, can inverses be defined. For any transformation T that is one-to-one and onto,

$$u \xrightarrow{T} v \qquad \text{or} \qquad v = T(u),$$

the inverse, denoted T^{-1}, is defined as the mapping obtained by reversing the sense of the T mapping,

$$v \xrightarrow{T^{-1}} u \qquad \text{or} \qquad u = T^{-1}(v).$$

That was explained in Section 2.2.

Under composition by consecutive application of transformations, T and T^{-1} are mutual inverses, since their compositions in both orders are the identity transformation,

$$T^{-1}T = TT^{-1} = I.$$

That is because

$$u \xrightarrow{T} v \xrightarrow{T^{-1}} u \qquad \text{or} \qquad (T^{-1}T)(u) = T^{-1}(T(u))$$

$$\underbrace{\phantom{u \xrightarrow{T} v}}_{T^{-1}T}$$

$$= I \qquad\qquad\qquad\qquad = T^{-1}(v)$$
$$= u$$
$$= I(u)$$

for all states u, and

$$v \xrightarrow{T^{-1}} u \xrightarrow{T} v \qquad \text{or} \qquad (TT^{-1})(v) = T(T^{-1}(v))$$

$$\underbrace{\phantom{v \xrightarrow{T^{-1}} u}}_{TT^{-1}}$$

$$= I \qquad\qquad\qquad\qquad = T(u)$$
$$= v$$
$$= I(v)$$

for all states v.

It should be clear by now that we are discovering a group. The set of all invertible transformations of any state space of a system forms a group under composition of consecutive application, called a *transformation group* of the system. (1) The set is closed under composition, since the composition of invertible transformations is an invertible transformation. Indeed, the composition of one-to-one (injective) transformations is a one-to-one transformation, and, if every state of a state space serves as an image for each of two transformations, every state will serve as an image for their composition [or, in short, the composition of onto (surjective) transformations is an onto transformation]. (2) We proved associativity. (3) The identity transformation is invertible. (4) Existence of inverses holds by definition. Note that a transformation group of a system is not unique, i.e., *the* transformation group of a system is not defined in general, as the same system might have different trasformation groups for different state spaces.

It should be pointed out that a transformation that is one-to-one and onto (bijective), i.e., invertible, is a permutation of state space. If such a transformation is represented by a two-column table, the list of states in the image column will be just a permutation of the list of states in the object column. Thus the transformation group of any finite state space containing n states is isomorphic with the symmetric group (not to be confused with symmetry group, which we study in Section 4.5) of degree n, S_n. For infinite state spaces the idea is the same, although perhaps not so clear intuitively.

Examples include the following.

1. Consider the equilibrium macrostates for the system of a given amount of a certain confined pure gas. Since p, V, θ (pressure, volume, absolute temperature) are related by an equation of state, we are free to vary any two of them, as long as p, V, $\theta > 0$ (assuming an ideal gas for simplicity). Thus all

transformations of the system are of the form

$$\begin{cases} p \to p' = f(p, \theta), \\ \theta \to \theta' = g(p, \theta), \end{cases}$$

for arbitrary positive functions $f(p, \theta)$, $g(p, \theta)$. For arbitrary invertible positive functions whose range of values is all the positive numbers we obtain the (infinite-order) transformation group of the system for its macrostate space.

2. Considering microstates for the system of a given amount of a certain pure gas, each molecule (all of which are identical, since the gas is pure) has three degrees of freedom (assuming spinless, structureless, point particles), so its state is characterized by three coordinates, x, y, z, and three components of momentum, p_x, p_y, p_z. All coordinates and all components of momentum for all molecules can be varied freely. Thus all transformations of the system are of the form

$$\begin{cases} x_i \to x_i' = X_i(x_j, y_j, z_j, p_{xj}, p_{yj}, p_{zj}), \\ y_i \to y_i' = Y_i(x_j, y_j, z_j, p_{xj}, p_{yj}, p_{zj}), \\ z_i \to z_i' = Z_i(x_j, y_j, z_j, p_{xj}, p_{yj}, p_{zj}), \\ p_{xi} \to p_{xi}' = P_{xi}(x_j, y_j, z_j, p_{xj}, p_{yj}, p_{zj}), \\ p_{yi} \to p_{yi}' = P_{yi}(x_j, y_j, z_j, p_{xj}, p_{yj}, p_{zj}), \\ p_{zi} \to p_{zi}' = P_{zi}(x_j, y_j, z_j, p_{xj}, p_{yj}, p_{zj}), \end{cases}$$

for $i, j = 1, \ldots, N$, where N is the number of molecules, with arbitrary functions $X_i(\ )$, $Y_i(\ )$, $Z_i(\ )$, $P_{xi}(\ )$, $P_{yi}(\ )$, $P_{zi}(\ )$. For arbitrary invertible functions whose range of values is all the real numbers we obtain the (infinite-order) transformation group of the system for its microstate space.

3. For a plane figure of given shape and size, lying in a given plane, with one of its points fixed in the plane, the angle of rotational orientation in the plane, γ, and its handedness can be varied freely. Thus all transformations of this system are

$$\gamma \to \gamma' = g(\gamma),$$

where the values of otherwise arbitrary $g(\gamma)$ are nonnegative and less than 360° [These are orientation-dependent rotations within the given plane about the fixed point by angle $g(\gamma) - \gamma$.] and reflections through all two-sided mirrors perpendicular to the given plane and passing through the fixed point. Restricting the functions $g(\gamma)$ to being invertible and having all nonnegative angle values less than 360°, we obtain, including the reflections, a transformation group (of infinite order) of the system. If, however, the transformation

$$\gamma \to \gamma' = g(\gamma)$$

is understood, not as giving the new orientation of the whole figure for every possible original orientation, but rather as giving the new orientation of each point of the figure for every possible original orientation of the point, whatever the state of the figure as a whole, as, in fact, is the usual interpretation (see the following section), then for the figure to retain its shape all its points must rotate through the same angle, and the functions $g(\gamma)$ are immediately restricted to the form

$$g(\gamma) = \gamma + \alpha.$$

4. For the system of one ball and three depressions the transformation group is the group of permutations of the three states of its state space, the symmetric group of degree 3, S_3 (of order 6).

5. For the system of three balls in three depressions the transformation group is the group of permutations of the six states of its state space, the symmetric group of degree 6, S_6 (of order 720).

6. For any quantum system the set of all transformations is all the operators on its Hilbert space of states. The subset of all invertible operators constitutes the transformation group. However, general operators are not considered for quantum systems, but only linear and antilinear operators. The transformation group is correspondingly the subset of all invertible linear and antilinear operators. See Section 4.7 for a more detailed treatment of quantum systems. [Recall that an antilinear operator A is characterized by the properties

$$A(|u\rangle + |v\rangle) = A|u\rangle + A|v\rangle,$$

$$Az|u\rangle = \bar{z}A|u\rangle$$

for all vectors $|u\rangle$ and $|v\rangle$ and all complex numbers z, where $\bar{z}$ is the complex conjugate of z.]

PROBLEM

Describe the transformation group for each state space of the problems of the preceding section.

4.3. Transformations in Space, Time, and Space–Time

Although we defined transformations in a very general way, and, as we saw in the examples, transformations can be very diverse indeed, transformations in space, in time, and in space–time are especially important, since the states of many systems are described in spatiotemporal terms. We therefore devote a section to a brief summary of some of the more important transformations of these kinds. We use Cartesian coordinates (x, y, z) (after René Descartes, French philosopher, mathematician, and scientist, 1596–1650) for space and Minkowskian coordinates (x, y, z, t) (Hermann Minkowski, Russian mathematician, 1864–1909) for space–time. First consider a number of spatial transformations.

The transformation of *spatial displacement* (or *spatial translation*) maps all points of space to image points that are the same distance away and in the same direction. Spatial displacements are invertible transformations.

$$\begin{cases} x \to x' = x + a, \\ y \to y' = y + b, \\ z \to z' = z + c, \end{cases} \qquad \begin{cases} x' \to x = x' - a, \\ y' \to y = y' - b, \\ z' \to z = z' - c. \end{cases}$$

The *rotation* transformation maps all points of space to image points found by rotating about a fixed common axis by a common angle. Rotations are invertible. If the axis of rotation is taken as the z-axis and the rotation angle is α (in the positive sense, from the positive x-axis toward the positive y-axis), we have

$$\begin{cases} x \to x' = x \cos \alpha - y \sin \alpha, \\ y \to y' = x \sin \alpha + y \cos \alpha, \\ z \to z' = z, \end{cases} \qquad \begin{cases} x' \to x = x' \cos \alpha + y' \sin \alpha, \\ y' \to y = -x' \sin \alpha + y' \cos \alpha, \\ z' \to z = z'. \end{cases}$$

See [M46] and [M45].

The *plane reflection* (or *plane inversion*) transformation is the transformation of reflection through a fixed two-sided plane mirror, the reflection plane. The image of any point is found by dropping a perpendicular from the point to the reflection plane and continuing the line on for the same distance on the opposite side of the plane. The image point is located at the end of that line segment. Plane reflections are invertible. If the reflection plane is taken as the xy-plane, we have

$$\begin{cases} x \to x' = x, \\ y \to y' = y, \\ z \to z' = -z, \end{cases} \qquad \begin{cases} x' \to x = x', \\ y' \to y = y', \\ z' \to z = -z'. \end{cases}$$

The transformation of *line inversion* (or *line reflection*) is inversion through a fixed straight line, the inversion line. The image of any point is the point at the end of the line segment constructed by dropping a perpendicular from the original point to the inversion line and continuing the perpendicular on for the same distance. Line inversions are invertible. If the inversion line is taken as the z-axis, we have

$$\begin{cases} x \to x' = -x, \\ y \to y' = -y, \\ z \to z' = z, \end{cases} \qquad \begin{cases} x' \to x = -x, \\ y' \to y = -y, \\ z' \to z = z. \end{cases}$$

By putting $\alpha = 180°$ in the rotation formulas above, we find that line inversion and rotation by $180°$ about the inversion line are the same transformation.

For the *point inversion* (or *point reflection* or *space inversion*) transformation the image of any point is at the end of the line segment running from the object point through a fixed point, the inversion center, and on for the same distance. Point inversions are invertible. If the inversion center is taken as the coordinate origin, we have

$$\begin{cases} x \to x' = -x, \\ y \to y' = -y, \\ z \to z' = -z, \end{cases} \qquad \begin{cases} x' \to x = -x', \\ y' \to y = -y', \\ z' \to z = -z'. \end{cases}$$

The *glide* transformation is the transformation consisting of the consecutive application of displacement parallel to a fixed plane and reflection through this plane, called a glide plane. (The two tranformations can just as well be applied in reverse order, since they commute.) Glide transformations are invertible. If the glide plane is taken as the xy-plane, we have

$$\begin{cases} x \to x' = x + a, \\ y \to y' = y + b, \\ z \to z' = -z, \end{cases} \qquad \begin{cases} x' \to x = x' - a, \\ y' \to y = y' - b, \\ z' \to z = -z'. \end{cases}$$

The *screw* transformation is the transformation resulting from the consecutive application (in either order, since they commute) of a rotation and a displacement parallel to the rotation axis, called a screw axis. Screw transformations are invertible. If the screw axis is taken as the z-axis, we have

$$\begin{cases} x \to x' = x \cos \alpha - y \sin \alpha, \\ y \to y' = x \sin \alpha + y \cos \alpha, \\ z \to z' = z + c, \end{cases} \qquad \begin{cases} x' \to x = x' \cos \alpha + y' \sin \alpha, \\ y' \to y = -x' \sin \alpha + y' \cos \alpha, \\ z' \to z = z' - c. \end{cases}$$

Under a *spatial dilation* (or *scale*) transformation the image of any point is found by moving the point away from a fixed point, the dilation center, along the straight line connecting them, to a distance from the fixed point related to the original distance by a fixed positive factor, the dilation factor. This transformation increases the distances between all pairs of points by the same factor. (If the dilation factor is less than 1, all distances are actually decreased.) Dilations are invertible. If the dilation center is taken as the coordinate origin, we have for positive ρ

$$\begin{cases} x \to x' = \rho x, \\ y \to y' = \rho y, \\ z \to z' = \rho z, \end{cases} \qquad \begin{cases} x' \to x = x'/\rho, \\ y' \to y = y'/\rho, \\ z' \to z = z'/\rho. \end{cases}$$

The image of any point under a *plane projection* transformation is a point in a fixed plane, the projection plane. It is the point of penetration of the perpendicular dropped from the original point to the projection plane. Projections are not invertible; they are neither one-to-one nor onto. If the projection plane is taken as the xy-plane, we have

$$\begin{cases} x \to x' = x, \\ y \to y' = y, \\ z \to z' = 0. \end{cases}$$

Under a *line projection* transformation the image of any point is a point on a fixed line, the projection line. The image is the point of intersection of the perpendicular dropped from the original point to the projection line. If the projection line is taken as the z-axis, we have

$$\begin{cases} x \rightarrow x' = 0, \\ y \rightarrow y' = 0, \\ z \rightarrow z' = z. \end{cases}$$

Now look at three temporal transformations. All are invertible. The first is *temporal* (or *time*) *displacement* (or *translation*), under which the images of all instants are the same time interval away from the original instants:

$$t \rightarrow t' = t + d, \qquad t' \rightarrow t = t' - d.$$

The transformation of *temporal* (or *time*) *inversion* (or *reflection* or *reversal*) maps each instant to the instant that is the same time interval before a fixed instant, the central instant, as the original instant is after it, or vice versa. If the inversion central instant is taken as the temporal origin, we have

$$t \rightarrow t' = -t, \qquad t' \rightarrow t = -t'.$$

The *temporal* (or *time*) *dilation* (or *scale*) transformation maps all instants to instants whose time intervals from a fixed instant, the central instant, are larger by a fixed positive factor, the dilation factor, than those of the respective object instants. This transformation increases the time interval between all pairs of instants by the same factor. (If the dilation factor is less than 1, all time intervals are actually decreased.) If the dilation central instant is taken as the temporal origin, we have for positive σ

$$t \rightarrow t' = \sigma t, \qquad t' \rightarrow t = t'/\sigma.$$

Our first spatiotemporal transformation is the *Lorentz transformation* (Hendrik Antoon Lorentz, Dutch physicist, 1853–1928), also called *boost*, or *velocity boost*, which maps all events to the events whose coordinates are the same as those that an observer moving with constant rectilinear velocity would assign to the original events with respect to his or her rest frame (the Minkowskian frame with respect to which he or she is at rest). Lorentz transformations are invertible. These transformations are an essential ingredient of Albert Einstein's (German–American physicist, 1879–1955) special theory of relativity. If the observer is moving in the negative x direction with velocity v, such that $-c < v < c$, where c is the speed of light, and if his or her Minkowskian coordinate axes are parallel to ours and his or her origin coincides with ours at time $t = 0 = t'$, we have

$$\begin{cases} x \rightarrow x' = \gamma(x + vt), \\ y \rightarrow y' = y, \\ z \rightarrow z' = z, \\ t \rightarrow t' = \gamma(t + vx/c^2), \end{cases} \qquad \begin{cases} x' \rightarrow x = \gamma(x' - vt'), \\ y' \rightarrow y = y', \\ z' \rightarrow z = z', \\ t' \rightarrow t = \gamma(t' - vx'/c^2), \end{cases}$$

where
$$\gamma = 1/\sqrt{1 - v^2/c^2}.$$

Our other spatiotemporal transformation is the *Galilei transformation* (Galileo Galilei, Italian astronomer, mathematician, and physicist, 1564–1642), also called *nonrelativistic boost* or *nonrelativistic velocity boost*, defined like the Lorentz transformation, except that the mathematical limit of $c \to \infty$ (or $v/c \to 0$) is taken. With the same assumptions as above we have

$$
\begin{cases}
x \to x' = x + vt, \\
y \to y' = y, \\
z \to z' = z, \\
t \to t' = t,
\end{cases}
\qquad
\begin{cases}
x' \to x = x' - vt', \\
y' \to y = y', \\
z' \to z = z', \\
t' \to t = t'.
\end{cases}
$$

PROBLEMS

1. Except for the projection transformations, all transformations and sets of transformations presented in this section generate or form groups. Every set of transformations containing continuous parameters forms a group, called a continuous group, where certain restrictions might apply to the ranges of the parameters. Prove that for each such set of transformations except the glide and find the restrictions, if any. Prove that each reflection or inversion transformation generates an order-2 group. (Groups of finite or denumerably infinite order are called discrete groups.) The set of glide transformations does not form a group, because it is not closed and the identity is missing. Prove, however, that the set of glide transformations with fixed glide plane generates a group.

2. Some of the transformations presented in this section can be generalized easily. Generalize the following transformations as indicated: (a) rotation about arbitrary axis parallel to the z-axis; (b) plane reflection with arbitrary reflection plane; (c) line inversion with arbitrary inversion line; (d) point inversion with arbitrary inversion center; (e) glide with arbitrary glide plane; (f) spatial dilation with arbitrary dilation center; (g) plane projection with arbitrary projection plane; (h) line projection with arbitrary projection line; (i) temporal inversion with arbitrary central instant; (j) temporal dilation with arbitrary central instant; (k) Lorentz transformation with arbitrary direction of velocity v; and (l) Galilei transformation with arbitrary direction of velocity v.

3. Many of the sets of transformations presented in this section can be combined and expanded to form larger groups than those discussed in Problem 1. That is especially simple when the transformations commute, such as spatial displacements parallel to the z-axis and rotations about that axis, generating the group of screw transformations. When the transformations do not commute, one must be careful. Prove that each of the following sets of transformations forms or generates a group: (a) rotations about the z-axis and spatial dilations (with center at the origin); (b) spatial displacements parallel to the xy-plane and rotations about all axes parallel to the z-axis; (c) spatial displacements and spatial dilations with all dilation centers; (d) spatial displacements, plane reflections with reflection plane parallel to the xy-plane, and glides with glide planes parallel to the xy-plane; (e) Lorentz

transformations with velocity v parallel to the x-axis, spatial displacements parallel to the x-axis, and temporal displacements; and (f) like (e) but with Galilei transformations instead of Lorentz transformations.

4.4. State Equivalence

We now consider the possibility of defining an equivalence relation for a state space of a system. That is, we recall (Section 2.4), any relation, denoted $\equiv$, that might hold between any two states and satisfies the following three conditions:

1. *Reflexivity.*

$$u \equiv u$$

for all states u.

2. *Symmetry*

$$u \equiv v \quad \Leftrightarrow \quad v \equiv u$$

for all states u, v.

3. *Transitivity.*

$$u \equiv v, \quad v \equiv w \quad \Rightarrow \quad u \equiv w$$

for all states u, v, w.

Recall that any subset of states of a state space—such a subset is called a subspace—such that all the states comprising it are equivalent to each other and no other state is equivalent to any state of the subset is called an equivalence class, or, in our case, an equivalence subspace. Recall also that an equivalence relation brings about a decomposition of a state space into equivalence subspaces. (See Section 3.1.)

Some examples of state equivalence follow.

1. If microstates of the gas of previous examples are being considered, while only the macroscopic properties of the gas are really of interest, any two microstates corresponding to the same pressure, volume, and temperature can be taken as equivalent. Actually there is an infinite number of microstates corresponding to any given macrostate and thus forming an equivalence subspace. As for the infinity of microstates that do not correspond to any equilibrium macrostate, we can define all of those as equivalent to each other.

2. Considering only macrostates of the gas, for some purposes states with the same temperature, for example, might be equivalent. That would occur if the gas served solely as a heat sink. Then all states with the same temperature would form an equivalence subspace.

3. If we take an equilateral triangle for the plane figure of given shape and size, lying in a given plane, with one of its points fixed in the plane, as far as appearance is concerned any state of the triangle will be equivalent to the two states obtained from it by rotations by 120° and by 240° about the axis

through its center and perpendicular to its plane and to the three states obtained from it by reflection through the three planes containing this axis and a median. Thus the state space of this system decomposes into equivalence subspaces of six equivalent states each.

4. If this triangle has one vertex marked or if it is deformed into an isosceles triangle, then the states obtained from any given state by rotations by 120° and 240° will no longer be equivalent to the original state with respect to appearance. And of the reflections only reflection through the plane containing the median of the different vertex and perpendicular to the plane of the triangle will still produce an equivalent state. Now the state space of the system decomposes into equivalence subspaces of only two states each.

5. If yet another vertex of the equilateral triangle is marked, and differently from the first, or if the isosceles triangle is further deformed into a scalene triangle, then all equivalence among states, with regard to appearance, will disappear.

6. In the system of one ball and three depressions, if the three depressions are adjacent to one another and the scene is viewed from a distance, as far as appearance is concerned all three states of the system will be equivalent.

7. If only two of the depressions are adjacent to one another and the third is separated from those two, as far as appearance is concerned two of the states will be equivalent to each other and the third will not be equivalent to either of those, so that state space will decompose into two equivalence subspaces, one containing two states and the other containing one state.

8. If the system is viewed from close by, no state will look like another, and there will be no equivalence with regard to appearance.

9. In the system of three balls in three depressions, if all three balls look the same, then as far as appearance is concerned all six states of the system will be equivalent to each other.

10. If two balls look alike and the third is distinct, each state of the system will be equivalent to the state obtained from it by interchanging the similar balls. Thus the state space of six states will decompose into three equivalence subspaces of two states each.

11. If all three balls look different, there will be no equivalence among states with regard to their appearance.

12. Since quantum systems are especially important, we discuss quantum state equivalence and its consequences separately in Section 4.7.

PROBLEM

For each of the problems of Section 4.1 impose the state-equivalence relation specified here on the state space(s) of the system, describe equivalence subspaces, and describe the decomposition of the state space(s) into equivalence subspaces: (a) In Problem 1 cards of the same numerical value are equivalent; (b) in Problem 2 discuss the equivalence relation imposed there; (c) in Problem 3 flag positions of the same height are equivalent; (d) in Problem 4 states that look the same are equivalent: (e) in Problem 5 the equivalence relation is parallelism; (f) in Problem 6 equivalent states have the same amount of HCl; (g) in Problems 7 and 8 equivalent states are states that would be the same if protons and neutrons were the same; and (h) in Problem 9 equivalent states have the same energy, ignoring spin-spin and spin-orbit interactions.

4.5. Symmetry Transformations, Symmetry Group

At this point some words for those on the "concept" track, those who have already read Chapters 8–10. We may observe that the formalism we are developing and our conceptual discussion of symmetry are converging. A change is represented by a transformation, while transformation groups represent families of changes. An equivalence relation for a state space of a system brings about a decomposition of the state space into equivalence subspaces, which is a classification, which implies analogy, which is a symmetry (see Section 8.4).

Now also for those on the "application" track, even without Chapters 8–10 the Chapter 1 definition of symmetry as immunity to a possible change gives: The symmetry is that any switching about of states that replaces states only by states equivalent to them then leaves unchanged the property of states that is their membership in an equivalence subspace. In simpler language: As long as the equivalence relation is not the trivial one that every state is equivalent only to itself, it is always possible to make a nontrivial change, a switch of states, that replaces states by equivalent states only. Such a switch operates within the individual equivalence subspaces and does not mix them up. Thus the property of states that they each belong to a certain equivalence subspace is immune to such a switch. That gives symmetry as the immunity of equivalence class membership to a possible switching about of states.

In the symmetry formalism we are developing, decomposition of a state space into nontrivial equivalence subspaces represents the aspect of the situation that is left unchanged by possible state switching changes. And state switching changes are represented by transformations. So symmetry is represented by the existence of transformations that leave equivalence subspaces invariant, i.e., transformations that map every state to an image state equivalent to the object state. Such a transformation is called a *symmetry transformation* for the equivalence relation. Thus the defining property of a symmetry transformation S is

$$u \xrightarrow{S} v \equiv u \qquad \text{or} \qquad S(u) = v \equiv u$$

for all states u.

The set of all invertible symmetry transformations of a state space of a system for an equivalence relation forms a group, a subgroup of the transformation group, called the *symmetry group* of the system for the equivalence relation. (Do not confuse symmetry group with symmetric group!) That is seen as follows.

1. Closure follows from transitivity of the equivalence relation, as either of the following diagrams prove:

$$u \equiv v = S_1(u),$$

$$v \equiv w = S_2(v) = S_2(S_1(u))$$

or

$$\stackrel{\text{def}}{=} (S_2 S_1)(u),$$

$$u \equiv w = (S_2 S_1)(u),$$

for all states u and v and symmetry transformations S_1 and S_2. Thus for all symmetry transformations S_1 and S_2 their composition $S_2 S_1$ is also a symmetry transformation, and by similar reasoning so is $S_1 S_2$.

2. Associativity holds for composition by consecutive application, as proved in Section 4.2.

3. The identity transformation is a symmetry transformation. That follows from reflexivity of the equivalence relation:

$$u \xrightarrow{\,I\,} u \equiv u \qquad \text{or} \qquad I(u) = u \equiv u$$

for all states u.

4. The inverse of any invertible symmetry transformation is also a symmetry transformation. If S is an invertible symmetry transformation, so that

$$u \xrightarrow{\,S\,} v \equiv u \qquad \text{or} \qquad S(u) = v \equiv u$$

for all states u, then by the symmetry property of the equivalence relation

$$v \xrightarrow{\,S^{-1}\,} u \equiv v \qquad \text{or} \qquad S^{-1}(v) = u \equiv v$$

for all states v. Thus S^{-1} is indeed a symmetry transformation.

That proves that the set of all invertible symmetry transformations of a state space of a system for an equivalence relation forms a group, a subgroup of the transformation group of the state space. In general a system might have different symmetry groups for its different state spaces and for the different equivalence relations that might be defined for them. If no two states are equivalent, i.e., if every state is equivalent only to itself, the symmetry group will consist only of the identity transformation, and the system will be asymmetric. In that case, although changes are possible, there are none that leave an aspect of the situation unchanged. In the other extreme, if all states are equivalent to each other so that all of state space is a single equivalence space, the symmetry group will be the transformation group itself.

That line of reasoning, by which any equivalence relation determines a subgroup of the transformation group, can be reversed to allow any subgroup of the transformation group to determine an equivalence relation. Given such a subgroup, the appropriate equivalence relation is simply: State

u is equivalent to state v if and only if some element T of the subgroup performs

$$u \xrightarrow{T} v \quad \text{or} \quad v = T(u).$$

That this relation is indeed an equivalence is seen as follows:

1. The reflexivity property follows from the identity transformation's belonging to the subgroup.
2. The symmetry property follows from the existence of inverses for the subgroup.
3. Transitivity follows from closure.

Such a subgroup might or might not be the symmetry group for the equivalence relation it determines in that way. Clearly all elements of the subgroup are symmetry transformations for the equivalence relation it determines. There might, however, be additional invertible symmetry transformations. If not, the subgroup is the symmetry group. In any case it is a subgroup of the symmetry group.

Consider some examples of symmetry groups.

1. In the case of the gas, where microstates corresponding to the same macrostate are considered equivalent to each other, the (infinite-order) symmetry group, transforming microstates into equivalent microstates only, is a subgroup of the (infinite-order) transformation group.

2. If macrostates of the gas having the same temperature are considered equivalent, we will have the constraint $\theta(p, V) =$ const. (isotherms), and all symmetry transformations will be of the form

$$p \rightarrow p' = f(p, \theta)$$

for arbitrary positive functions $f(p, \theta)$, a subset of all transformations of this state space as presented in Section 4.2. For arbitrary invertible positive functions $f(p, \theta)$ whose range of values is all positive numbers we obtain the (infinite-order) symmetry group, a subgroup of the (infinite-order) transformation group.

3. The symmetry group of the equilateral triangle discussed in Section 4.4 is the group consisting of the identity transformation, rotations about the axis through the center of the triangle and perpendicular to its plane by 120° and 240°, and reflections through each of the three planes containing the rotation axis and a median, the order-6 group D_3. Refer to Fig. 4.1. That group is a subgroup of the transformation group of the system, which is the same as the transformation group of any plane figure of given shape and size, lying in a given plane, with one of its points fixed in the plane, and was presented in Section 4.2.

4. For the equilateral triangle with one vertex marked or for an isosceles triangle the symmetry group consists only of the identity transformation and reflection through the single plane containing the median of the distinct vertex and perpendicular to the plane of the triangle, the order-2 group C_2. See

Fig. 4.1. Equilateral triangle. Center is point of intersection of axis of threefold rotation symmetry. Medians are lines of intersection of planes of reflection symmetry.

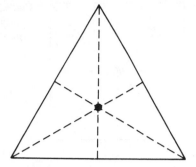

Fig. 4.2. That group is a subgroup of the transformation group of the system and is also a proper subgroup of the symmetry group of the equilateral triangle.

5. For the equilateral triangle with two vertices marked differently or for a scalene triangle the symmetry group consists only of the identity transformation. Those systems are asymmetric. This trivial symmetry group is a subgroup of the transformation group of the system and is also a proper subgroup of the symmetry group of the singly marked equilateral triangle or the isosceles triangle and thus also a proper subgroup of the symmetry group of the equilateral triangle.

6. For a similar example, replace the equilateral triangle of the preceding examples by a square. The symmetry group of that system for the equivalence relation of identical appearance consists of the identity transformation, rotations by 90°, 180°, and 270° about the axis through the center of the square and perpendicular to its plane, reflections through each of the two planes containing the axis and a diagonal, and reflections through each of the two planes containing the axis and parallel to a pair of edges, the order-8 group D_4. Refer to Fig. 4.3. That group is a subgroup of the transformation group of the system (the same as that of any plane figure of given shape, etc.).

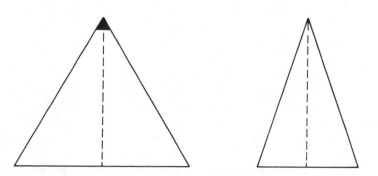

Fig. 4.2. Equilateral triangle with one vertex marked and isosceles triangle. Median of distinct vertex is line of intersection of plane of reflection symmetry.

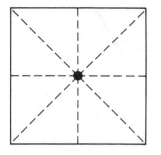

Fig. 4.3. Square. Lines of intersection of planes of reflection symmetry are indicated. Center is point of intersection of fourfold rotation symmetry axis.

We take this opportunity to give the definition of the general *dihedral group* D_n. In generalization of the symmetry groups of the equilateral triangle and the square, the dihedral group D_n is the symmetry group of the regular *n*-sided plane polygon. It consists of the identity transformation, rotations about the axis through the center of the polygon and perpendicular to its plane by $360°/n$, $2 \times 360°/n$, $3 \times 360°/n$, ..., $(n - 1) \times 360°/n$, and reflections through each of the *n* planes containing the axis and a vertex or the center of a side (or both). D_n is of order $2n$.

7. If the square is squeezed into a rectangle, its symmetry group will reduce to the identity transformation, rotation by 180° about the axis through the center of the rectangle and perpendicular to its plane (or inversion through the axis, which is the same), and reflections through each of two planes containing the axis and parallel to a pair of edges, the order-4 group D_2. Refer to Fig. 4.4 That is a subgroup of the transformation group of the system (that of any plane figure of given shape, etc.) and is also a proper subgroup of the symmetry group of the square.

8. If in the system of one ball and three depressions all three depressions are adjacent so that all three states look the same from a distance, then all states will be equivalent, and the symmetry group will be the transformation group itself, the order-6 group S_3. The symmetry group in that case happens to be isomorphic with the symmetry group of the equilateral triangle of given size, etc., D_3.

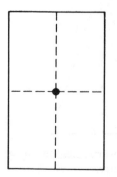

Fig. 4.4. Rectangle. Lines of intersection of planes of reflection symmetry are indicated. Center is point of intersection of twofold rotation symmetry axis.

9. If only two of the depressions are adjacent, so that only two of the three states are equivalent to each other, the symmetry group will consist of the identity transformation and the transposition interchanging the two equivalent states, the order-2 group S_2. That group is a subgroup of the transformation group of the system, S_3, which is also the symmetry group of the system with all depressions adjacent. In addition, S_2 is isomorphic with the symmetry group of the isosceles triangle (or equilateral triangle with one vertex marked) of given size, etc., C_2.

10. If all three depressions are separated and no states are equivalent, the symmetry group will consist only of the identity transformation. That is a subgroup of the transformation group and is also a proper subgroup of the symmetry group of the system with two adjacent depressions and thus also a proper subgroup of the symmetry group of the system with three adjacent depressions.

11. If in the system of three balls in three depressions all three balls look the same, then, since all states are equivalent, the symmetry group will be the transformation group itself, the order-720 group S_6.

12. If only two balls look alike and the third is distinct, the symmetry group will be the direct product of the permutation groups of the individual equivalence subspaces, the direct product of the symmetric group of degree 2 with itself twice, $S_2 \times S_2 \times S_2$, an order-8 group. That group is a subgroup of the transformation group of the system, S_6, which is also the symmetry group of the system with three similar balls.

13. If all three balls look different, the symmetry group will consist only of the identity transformation. That is a subgroup of the transformation group and is also a proper subgroup of the symmetry group of the system with two similar balls and thus also a proper subgroup of the symmetry group of the system with three similar balls.

Note that in each of examples 8 to 13 the state space is finite and we found the symmetry group exactly according to its definition. In each of examples 1 to 7, however, the state space is infinite and decomposes into an infinite number of equivalence subspaces. What we found is the transformation group of each equivalence subspace rather than the symmetry group of the entire state space. The latter is the infinite direct product of the transformation groups of the individual equivalence subspaces, which is so awkward that the normal procedure is to do just what we did.

14. Symmetry groups for quantum systems are discussed in Section 4.7.

PROBLEMS

1. Prove in detail that a subgroup of the transformation group of a state space defines an equivalence relation for the state space, as described in this section, and that the defining subgroup is a subgroup of the symmetry group for the equivalence relation it defines.

2. Prove that the symmetry groups of the equilateral triangle and square, of given size and given location in a given plane, with identical-looking states considered equivalent, are indeed D_3 and D_4, respectively, as claimed in the text.

3. Prove that the symmetry group of the system of three balls in three depressions with two of the balls similar is $S_2 \times S_2 \times S_2$, as claimed in example 12.

4. For each of the Platonic solids find all the rotations, plane reflections, and point inversions that are symmetry transformations. (It is a very good idea to construct models and work with them.) Compare your results for the cube with those for the regular octahedron and your results for the regular dodecahedron with those for the regular icosahedron. In what way is the regular tetrahedron "odd man out"?

5. Find all the geometric symmetry (i.e., appearance-preserving) transformations of an infinite bathroom floor tiled with identical regular hexagons.

6. Find all the symmetry (i.e., appearance-preserving) transformations for (a) a standard chess board and (b) an infinite chess board. (*Hint:* In addition to geometric transformations consider the transformation of black–white color interchange and its composition with other transformations.)

7. Find the symmetry group for each state space of the problems of Section 4.1 with the state-equivalence relation specified in the problem of Section 4.4.

8. As an example of the interesting byways into which symmetry can lead us, a person found that both $S_2 \times S_2 \times S_2 \times S_2 \times S_2 \times S_2 \times S_2 \times S_2$ and $S_9 \times S_9 \times S_2$ were symmetry groups of the 20 horses in his or her stable. What equivalence relations might he or she have had in mind?

4.6. Quantification of Symmetry

We now have at our disposal the means to attempt to quantify symmetry or at least set up a hierarchy, or ordering, or symmetries. A quantification of symmetry would be a way of assigning a number to each symmetry group, expressing the degree of symmetry of a system having that symmetry group. A symmetry ordering would be a way of comparing any two symmetry groups to determine which of two systems, each possessing respectively one of the two symmetry groups, has the higher degree of symmetry. A symmetry quantification is clearly a symmetry ordering, although an ordering might not go so far as to be a quantification.

There are three properties that, on the basis of our experience, we may reasonably expect of any scheme of symmetry quantification or ordering. First of all, if the symmetry group of a system is isomorphic with the symmetry group of another system (e.g., the equilateral triangle of given size, etc., and the ball and three adjacent depressions, or the isosceles triangle and the ball and two adjacent and one separated depressions), it is reasonable to consider them as having the same degree of symmetry (even though their symmetry transformations might be of entirely different character, as in the examples). Thus it is the abstract group of which the symmetry group is a realization that is of interest, rather than the symmetry group itself.

Next, if the symmetry group of a system is isomorphic with a proper subgroup of the symmetry group of another system (e.g., the isosceles triangle and the equilateral triangle, or the ball and two adjacent and one separated depressions and the ball and three adjacent depressions, or the isosceles triangle and the ball and three adjacent depressions, or the rectangle and the square), the latter system many reasonably be considered more symmetric than the former.

And finally, if a system is asymmetric—if its symmetry group consists solely of the identity transformation—it is reasonable to assign it the lowest degree of symmetry.

But how should we compare the degrees of symmetry of systems whose symmetry groups are neither isomorphic with each other nor one isomorphic with a proper subgroup of the other (e.g., the square and the equilateral triangle)? The main purpose of any symmetry quantification or ordering scheme is to answer just that question.

One possibility for a quantification scheme is to take for the degree of symmetry of a system the order of its symmetry group (or any monotonically ascending function thereof, such as its logarithm). For finite-order symmetry groups (and we do not go into infinite-order groups here) that is a perfectly satisfactory scheme in that it possesses the three properties we demand previously: Isomorphic groups have the same order, a proper subgroup is of lower order than its including group, and the order-1 group has the lowest order of all. It might, however, well be objected on philosophical grounds that this quantification scheme assigns equal weights to all the symmetry transformations of a symmetry group, while in fact they are not all independent, as they can be generated by repeated and consecutive applications of a minimal set of generators. We are thus led to consider the scheme whereby the degree of symmetry of a system is taken to be the minimal number of generators of its symmetry group. As for the three properties, it possesses the first; isomorphic groups have the same minimal numbers of generators. It is, however, deficient with respect to the second property; a proper subgroup does not necessarily have a lesser minimal number of generators than its including group. For example, the order-2 cyclic group C_2 is a proper subgroup of the order-4 cyclic group C_4, but, being cyclic, each group can be generated by positive powers of a single element. The third property can be forced by assigning the value 0 to the order-1 group.

For comparison we list in Fig. 4.5 some of the symmetries we found in our examples along with the degrees of symmetry assigned to the systems by the two symmetry-quantification schemes considered. Note, for example, that the minimal-number-of-generators scheme assigns equal degrees of symmetry to the square and to the rectangle. The group-order scheme is clearly preferable. Other, intermediate schemes could be devised, giving the degree of

System	Symmetry group	Order	Minimal number of generators
Square	D_4	8	2
Rectangle	D_2	4	2
Equilateral triangle	$D_3 \sim S_3$	6	2
Isosceles triangle	$C_2 \sim S_2$	2	1
Ball and three adjacent depressions	$S_3 \sim D_3$	6	2
Ball and two adjacent and one separated depressions	$S_2 \sim C_2$	2	1
Asymmetric system	C_1	1	0 (defined)

Fig. 4.5. Table of orders and minimal numbers of generators for various symmetry groups.

symmetry of a system as some function of both the order and the minimal number of generators of its symmetry group.

4.7. Quantum Systems

This section is a concise presentation intended to place the subject of symmetry in quantum systems within the formalism developed in the preceding sections of this chapter. Thus it should introduce, but by no means replace, the various discussions of the same subject found elsewhere. It is intended for readers with a good understanding of the Hilbert-space formulation of quantum theory; readers lacking that prerequisite are advised to skip it.

Quantum systems, which are but special cases of the general systems we have been discussing, are so important they deserve separate treatment. Their special nature allows us to be more specific in our discussion of their symmetry.

The state spaces of quantum systems are Hilbert spaces, and quantum states are vectors in those spaces. (Actually, quantum states are *represented* by vectors in Hilbert spaces. More accurately, they are represented by *rays*, since all complex multiples of any vector are supposed to represent the same state. But for the convenience of the present discussion we identify state with vector. For the sake of simplicity we also ignore possible complications due to coherent subspaces, which in any case can be overcome with slight additional effort.) The transformations of interest for quantum systems are linear and antilinear transformations in their Hilbert state spaces, said to be implemented by linear and antilinear operators. In the following discussion we use "operator" to mean "linear or antilinear operator." Thus the transformation group of a quantum system is the group of all invertible operators in its Hilbert space of states.

Now what properties of states of quantum systems might be utilized for the definition of equivalence relations? Consider the following:

1. *Transition amplitudes.* Each state $|u\rangle$ is characterized by the transition amplitudes between itself and all states of its Hilbert space, $\langle u|v\rangle$ or $\langle v|u\rangle$ for all states $|v\rangle$.

2. *Norm.* Especially, each state $|u\rangle$ is characterized by its transition amplitude with itself, $\langle u|u\rangle$, the square of its norm.

3. *Eigenheit.* Each state $|u\rangle$ either is or is not an eigenstate of each member of any set of operators $\{O_i\}$ in its Hilbert space, and, if it is, it is characterized by its eigenvalue.

4. *Expectation values.* Each state $|u\rangle$ gives a set of expectation values $\{\langle u|O_i|u\rangle/\langle u|u\rangle\}$ for any set of operators $\{O_i\}$ in its Hilbert space.

It seems reasonable to define state equivalence for a given quantum system and a given set of operators in its Hilbert state space as indistinguishability with respect to properties 1 through 4.

The search for all symmetry operators for the equivalence relation of indistinguishability with respect to (1) transition amplitudes and (2) norm is essentially reducible to the search for all operators preserving scalar products up to complex conjugation. And the standard result is the group of all unitary and antiunitary operators in the Hilbert state space of the system. See [A1]. For strict preservation of scalar products the symmetry groups consists only of all unitary operators. [Recall that antiunitary (or unitary antilinear) operators are invertible antilinear operators A obeying

$$\langle u|A^\dagger A|v\rangle = \langle v|u\rangle = \overline{\langle u|v\rangle}$$

for states $|u\rangle$, $|v\rangle$, where the bar denotes complex conjugation and the dagger denotes Hermitian conjugation.]

Now consider property 3. Equivalent states, in addition to their being related by unitary or antiunitary operators, are also required to be either all eigenstates or all not eigenstates of each operator O_i of a given set $\{O_i\}$, and, if they are all eigenstates, they must belong to the same eigenvalue. The first part of this requirement leaves as symmetry operators only those unitary and antiunitary operators commuting with the set $\{O_i\}$. The second part in general rejects the antiunitary operators among those, since, while eigenstates related by unitary operators indeed belong to the same eigenvalue of an operator commuting with the unitary operators, a pair of eigenstates related by an antiunitary operator commuting with the operator of which they are eigenstates belong to eigenvalues that are a complex-conjugate pair. However, if the set of operators $\{O_i\}$ consists solely of Hermitian operators, which indeed is the usual case, with Hermitian operators representing measurable physical quantities, the antiunitary operators are *not* rejected. In that case a pair of eigenstates related by an antiunitary operator commuting with the Hermitian operator of which they are eigenstates belong to the same eigenvalue, since the eigenvalues of Hermitian operators are real. A very important Hermitian operators that is almost always included in the set $\{O_i\}$ is the Hamiltonian operator (after William Rowan Hamilton, British mathematician, 1805–1865), representing the energy of the system and generating the system's evolution.

That also takes care of property 4. States related by unitary operators commuting with a given set of operators $\{O_i\}$ give equal sets of expectation values. A pair of states related by an antiunitary operator commuting with the set give complex-conjugate sets of expectation values. But if $\{O_i\}$ is a set of Hermitian operators, they will give equal sets of (real) expectation values.

In summary, for a given quantum system and a given set of operators in its Hilbert space of states the symmetry group for equivalence with respect to (1) transition amplitudes, (2) norm, (3) eigenheit, and (4) expectation values is the group of all unitary operators in the Hilbert space commuting with the given set of operators. If the given set consists solely of Hermitian operators, the symmetry group is the group of all unitary and antiunitary operators commuting with the set. (But if it is desired to preserve scalar products strictly, not just up to complex conjugation, then the antiunitary operators are excluded whether the operators of the given set are Hermitian or not.)

PROBLEMS

1. Find the symmetry transformations of the harmonic oscillator, where the set $\{O_i\}$ consists of the Hamiltonian operator only.

2. Find the symmetry transformations of the nonrelativistic, spinless hydrogen atom, where the set $\{O_i\}$ consists only of the Hamiltonian operator.

3. Find the symmetry transformations of the nonrelativistic, spinless hydrogen atom in a uniform magnetic field, where the set $\{O_i\}$ consists only of the Hamiltonian operator.

4. Find the symmetry transformations of a relativistic free particle, where the set $\{O_i\}$ consists of the mass and spin operators.

4.8. Summary of Chapter Four

In this chapter we started to develop the general symmetry formalism needed for the application of symmetry considerations in science, especially quantitative applications. In Section 4.1 we introduced the following very general concepts: system, which is whatever we investigate the properties of; subsystem, a system wholly contained within a system; state of a system, a possible condition of the system; and state space of a system, which is the set of all states of the same kind. The concept of transformation, a mapping of a state space of a system into itself, was presented in Section 4.2. We saw that the set of all invertible transformations of a state space of a system forms a group, called a transformation group of the system.

Section 4.3 was a compilation of a number of transformations in space, in time, and in space–time. The spatial transformations presented were: displacement, rotation, plane reflection, line inversion, point inversion, glide, screw, dilation, plane projection, and line projection. The temporal transformations were displacement, inversion, and dilation. And the spatio-temporal transformations we listed were the Lorentz transformation and the Galilei transformation.

In Section 4.4 we considered the possibility of an equivalence relation for a state space of a system. Such a relation decomposes a state space into equivalence subspaces. That led in Section 4.5 to the idea of a symmetry transformation, which is any transformation that maps every state to an image state equivalent to the object state, i.e., any transformation that preserves equivalence subspaces. The set of all invertible symmetry transformations of a state space for an equivalence relation forms the symmetry group of that state space for that equivalence relation, a subgroup of the transformation group.

We discussed quantification of symmetry in Section 4.6, where we found that the order of a (finite-order) symmetry group, or any monotonically increasing function thereof, can reasonably serve as the degree of symmetry of a system possessing that symmetry group.

Our discussion of state equivalence for quantum systems in Section 4.7 led to the result that for a given quantum system and a given set of operators in its Hilbert space of states, the symmetry group is the group of all unitary operators in that Hilbert space commuting with the given set. If the given set consists solely of Hermitian operators, the symmetry group is the group of all unitary and antiunitary operators commuting with the set.

CHAPTER FIVE

Application of Symmetry

In this chapter we continue developing the general symmetry formalism that we started developing in the preceding chapter. Here we study the theory and practice of the application of symmetry in science. The practice is based on the symmetry principle, stating roughly that the effect is at least as symmetric as the cause, and the theory is the proof of that principle. The proof starts with a discussion of the concept of causal relation, followed by a clarification of certain points concerning the nature of science, from which we are led to the equivalence principle, stating roughly that equivalent causes imply equivalent effects. The symmetry principle follows almost immediately from the equivalence principle. In the application of symmetry in science the symmetry principle can be used in two ways: minimalistically, to set a lower bound on the symmetry of the effect, and maximalistically, to set an upper bound on the symmetry of the cause. Numerous examples are presented. The application of symmetry in quantum systems is briefly discussed.

Except for the concepts taken from the previous chapters, the material is developed from scratch. The theory leading to the equivalence principle is unavoidably philosophical, so be forewarned. If philosophical considerations are anathema to you, you had better just memorize the equivalence principle and skip the first two sections. But why not give it a try? It might not be as bad as you fear. However, Section 5.6, Quantum Systems, requires an understanding of quantum theory and its Hilbert-space formulation. If that is lacking, the section should be skipped. It is very brief anyhow. For the examples and problems the mathematical prerequisites of the previous chapters certainly suffice, except for a few cases that require an understanding of simple ordinary differential equations. The scientific prerequisite is an elementary understanding of certain topics in classical and modern physics: mechanics, electromagnetism, quantum mechanics, special relativity, and the Rutherford experiment (Ernest Rutherford, British physicist, 1871–1937).

5.1. Causal Relation

The first step in our development of the theory of application of symmetry is to achieve a sufficient understanding of the concept of *causal relation*, also called *cause–effect relation*. It might seem that in everyday affairs there should be no difficulty in understanding that relation; the cause brings about the effect, and the effect is a result of the cause. But is the situation really so clear? For example, at the end of a concert you applaud, whereby your clapping makes a noise that expresses your appreciation. The clapping makes the noise, so the clapping is the cause and the noise is the effect, which is satisfactory. But why do you clap? You clap to make the noise; if clapping did not make that kind of noise, you would not clap. Thus your clapping is performed because of the noise. And does that not mean that the noise is the cause and the clapping the effect? Or is it the desire to produce the noise that is the cause?

Possibly in everyday affairs we can make do with such foggy concepts, but in science our concepts must be much clearer (that would do no harm in everyday affairs either!). So we will attempt to clarify the concept of causal relation to a degree of hairsplitting sufficient for our purposes.

Let us recall that a subsystem of a system is itself a system, with states, state spaces, transformation groups, and perhaps even equivalence relations and symmetry groups. The state of a system determines the state of each of its subsystems. However, the state of a subsystem does not in general determine the state of the whole system or of any other subsystem (although that might happen in certain cases).

Consider two subsystems, which we denote A and B, of an arbitrary system. Consider all states of the whole system and imagine that we make a triple list of the pairs of states of subsystems A and B that are determined by each state of the whole system for all of its states. Of course, it is possible that the same state of a subsystem will appear more than once in the resulting list. See Fig. 5.1.

States of whole system	Determine states of	
	Subsystem A	Subsystem B
a	h	p
b	i	p
c	h	q
d	j	r
e	j	s
$\vdots$	$\vdots$	$\vdots$

Fig. 5.1. States of two subsystems determined by state of whole system.

States of whole system	Determine states of	
	Subsystem A	Subsystem B
a	h	p
b	i	q
c	j	r
d	i	q
e	l	r
f	m	s
⋮	⋮	⋮

Fig. 5.2. *A* is cause subsystem and *B* is effect subsystem.

Let us now look for a correlation between states of subsystem *A* and states of subsystem *B* in that list by asking the following questions: Does the same state of subsystem *A* always appear with the same state of subsystem *B* (and possibly different states of *A* appear with the same state of *B*)? Does the same state of subsystem *B* always appear with the same state of subsystem *A* (and possibly different states of *B* appear with the same state of *A*)? If the answer to either or both questions is affirmative, we say that a *causal relation* (or *cause–effect relation*) exists between the two subsystems. If the first question is answered affirmatively, we say that subsystem *A* is a *cause subsystem* and subsystem *B* is an *effect subsystem*. If the second question is answered affirmatively, we say that *B* is a cause subsystem and *A* is an effect subsystem. If the answers to both questions are affirmative, each subsystem is both cause and effect. See Figs. 5.2 and 5.3.

In that manner the definition of a causal relation between subsystems is

States of whole system	Determine states of	
	Subsystem A	Subsystem B
a	h	p
b	i	q
c	j	r
d	k	s
e	h	p
f	l	t
⋮	⋮	⋮

Fig. 5.3. Each subsystem in both cause and effect subsystem.

based on a correlation between their state spaces. There is not even a hint of "bringing about," "resulting," "producing," or "causing" (in the usual sense). The states of one subsystem and those of the other are connected through the fact that they are determined by the states of the same whole system, and that connection is the origin of possible causal relations between subsystems. If we think we understand the "mechanism" that underlines the correlation, we express the relation in terms of "producing," etc. But even if we do not understand the "mechanism" or perceive any "mechanism" at all, a causal relation may still exist between systems as an empirical fact of correlation between their state spaces. Upon discovering a pair of causally related systems, we indeed do tend to search for a whole system of which the causally related systems are subsystems, in order to "understand" the causal relation. However, the existence of a causal relation and our knowledge of its existence in no way depend on our understanding or perceiving its underlying "mechanism."

And how does that abstract definition relate to our intuitive understanding of the concept of causal relation? First of all, I would like to warn against attaching too much importance to one's intuitions, at least in science. Intuitions are no more than thought habits. And since those habits developed as a result of limited experience, their appropriateness to phenomena lying outside that range of experience is suspect, at the least. (Need I remind you of the history of the theories of special relativity and quantum mechanics?) Second of all, in spite of that there is no contradiction between our abstract definition of causal relation and our intuitive understanding of the concept. We are used to the idea that, if a cause produces an effect, every time the cause is in the same state the effect is in the same state. Otherwise we would not have even considered the "cause" as being a cause to begin with.

For example, we are used to thinking that forces produce acceleration, that the forces are the cause and the acceleration is the effect, and not vice versa. And indeed, whenever the same set of forces acts on a given body, the same acceleration occurs. We cannot agree that the acceleration is the cause and the forces the effect, because the same acceleration may be correlated with different sets of forces (having the same resultant). Different sets of forces may be correlated with the same acceleration, while different accelerations are never correlated with the same set of forces. In abstract terms the system is the body, the forces acting on it, and its acceleration. The forces are subsystem A and the acceleration is subsystem B. The answer to the first question is affirmative; the same state of forces is always correlated with the same state of acceleration, and even different states of forces are correlated with the same state of acceleration. The answer to the second question is negative; the same state of acceleration is not always correlated with the same state of forces. Therefore the forces and the acceleration are in causal relation, with the forces the cause and the acceleration the effect.

Let us return to our example of applause. The system is the human body, all its actions, and all its noises. The hands are subsystem A, which is considered to have only two states: clapping and not clapping. Subsystem B consists of all body noises and is also considered to have only two states: clapping noise and not clapping noise (i.e., silence or any other noise). As we review all states of the whole system, we find a most astounding correlation between the states of A and the states of B: Whenever the hands clap, clapping noise sounds; and whenever clapping noise sounds, the hands are clapping. Thus the answers to both the first and the second questions are affirmative, and there exists a causal relation between hand clapping and clapping noise, where each may be taken as cause, as effect, or as both.

Another example is the spatiotemporal configuration of electric charge and current densities as the cause and the electric and magnetic field strengths as effect. Or we could have the scattering potential and incident wave function as cause and the scattered wave function as effect. The cause might be your present situation and recent history, with your present mood as effect. And so on.

Or consider the famous Rutherford experiment, in which alpha particies are scattered from stationary metal nuclei. If the exact locations of the target nuclei and the precise trajectory of each incident particle could be known, those would be the cause, and the effect would be the point where each scattered alpha particle hits the fluorescent detection screen. However, the locations of the target nuclei and the trajectory of each incident particle are not known. The best we can do is to confine the metal nuclei to a thin foil of known density and to know the statistical properties of a beam of alpha particles (particle flux, velocity distribution, etc.). Taking these as the cause, the effect certainly cannot be the final trajectory of a scattered particle or even its scattering angle, since scattering in all directions is actually observed in the experiment. The effect is, in fact, the statistical distribution of scattering angles for the scattered alpha particles, expressed by the angular-distribution density function or differential cross section.

PROBLEMS

1. Discuss a possible causal relation between day and night both in terms of empirical correlation and in terms of mechanism.

2. Consider a possible causal relation between thirst and drink.

3. Consider a possible causal relation between crime and punishment.

4. Consider a possible causal relation between life and death.

5. One hears the claim that the existence of some attribute or quality implies the existence of its opposite. For example, the existence of good implies the existence of evil. The philosophically inclined reader might consider this claim from the point of view of causal relation.

5.2. The Equivalence Principle

The ultimate goal of our line of reasoning, which started with a discussion of the causal relation, is *the symmetry principle*. Along that line we must stop at two way stations, which cannot be avoided if we want not only to use the symmetry principle but to understand it. At the first station we must clarify to ourselves certain points concerning the nature of science. At the second stop we will show that the very existence of science implies *the equivalence principle*: *Equivalent causes–equivalent effects*. Those two issues are dealt with in the present section. In the following section the symmetry principle is derived as an almost immediate consequence of the equivalence principle, and we defer its formulation to there.

One ingredient of the foundation of science is reproducibility (about which see Sections 9.1 and 9.6). Reproducibility is the possibility of repeating experiments under conditions that are similar in certain respects, yet different in other respects, such as location or time, and obtaining essentially the same results. (Section 9.6 goes into detail.) That possibility furnishes the objectivity that is essential for science to be a common, lasting endeavor rather than a set of private sciences or impossible altogether. We are not claiming that all the phenomena in the world fulfill the condition of reproducibility. For instance, the phenomena of parapsychology are notorious for their irreproducibility. That fact does not negate the possible existence of such phenomena nor does it necessarily invalidate their investigation. But as long as reproducibility is not achieved in parapsychology, the latter will continue to lie outside the domain of concern of science.

Another ingredient of the foundation of science is predictability (about which see Sections 9.1 and 9.7). Predictability is the possibility of predicting the result of as yet unperformed experiments. This means that, until proved otherwise, we labor under the assumption that human intelligence is capable of understanding nature sufficiently to allow the prediction of phenomena that have not yet been observed. Reproducibility alone, without predictability, does not make science. It only allows recording, cataloging, and classifying of experimental observations as public information, with no benefit beyond the compilation itself. Science begins to be possible only when order is perceived among the collected facts, and on the basis of that order the results of new experiments are predicted. (See Section 9.7 for more detail.)

The scientific tool by which we predict the results of experiments is the law. A law is any conceptual recipe or mathematical formula or other means like those that, when fed data about the conditions of an experiment, gives us the experimental result to be expected. The first test of a proposed law is performed by comparing it with past experiments and their results. If the results expected according to the law match the results actually obtained, the law advances to the next testing stage. If not, it goes back to the drawing board for corrections or overhaul or into the wastebasket. For the next test the law must predict the results of as yet unperformed experiments. If it does that with continuing success, our confidence in it continually increases. However, a single failure is sufficient to invalidate a propsed law in spite of its past success.

Reproducibility and predictability are expressions or our realization that causal relations exist in nature. A scientific law is the expression of a particular causal relation; the data that the law receives represent the physical cause, and the results that the law gives represent the physical effect. When a law receives the same data, it always gives the same results, and that reflects the relation between the cause and the effect.

A scientific law may be considered as including a sort of antianalog computer. We use the term "antianalog computer," since, whereas analog computers represent mathematical procedures by physical processes, scientific laws lead to representations of physical processes by (almost always) mathematical procedures. This antianalog computer is equipped with a set of terminals into which input may be fed and out of which output may be obtained. Different subsets of terminals might be used to feed and obtain different types of input and output. Some terminals might be exclusively either for input or for output, while others might serve for both input and output (though not during the same application). Every input uniquely determines an output. (It is possible, of course, that the same output might be the result of different inputs.) That gives us reproducibility and predictability. A scientific law must also supply a set of rules for translating a physical situation into input acceptable by the antianalog computer (and even for deciding whether a physical situation can be so translated) and for translating the output into physical terms.

As an example, consider the laws of classical vacuum electromagnetism with Maxwell's equations (James Clerk Maxwell, Scottish physicist, 1831–1879) and certain boundary conditions as their antianalog computer. The input might be functions, as a mathematical translation of a physical situation in terms of electric charge and current densities. The output is then functions, which are translated as electric and magnetic field strengths, giving the forces on test charges.

As another example, consider the theory of nonrelativistic quantum mechanics with the Schrödinger equation (Erwin Schrödinger, Austrian physicist, 1887–1961) and certain boundary conditions as its antianalog computer. Our input often consists of functions, as a mathematical translation of a physical situation in terms of potential and incident wave function. The output is then a function, which is translated as the scattered wave function and from which are derived probability, cross sections, and such measurable quantities.

Consider a physical system and a law of its behavior. A system must have at least one pair of subsystems in causal relation for a law of its behavior to be possible, since, as stated above, a law is an expression of a causal relation in a system. Consider any such pair, a cause subsystem and an effect sub-

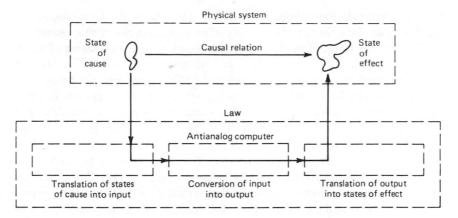

Fig. 5.4. Relation of law to cause and effect in physical system.

system. Since the law describes the behavior of the whole system, it must take the translation of the states of the cause subsystem as input to its antianalog computer, and the translation of the output of its antianalog computer must, by exhausting its total information content, determine states of the effect subsystem. Moreover, the state of the effect that is causally correlated with a given state of the cause must be consistent with the translation of the output that is uniquely produced by the antianalog computer as a result of receiving as input the translation of the given state of the cause. See Fig. 5.4.

States of causes and effects are actual physical situations occurring in actual physical systems. Each such state is unique; no two are identical. Although similar in many respects, a pair of states might still differ, for example, in their geographical locations, in their times of occurrence, or in the experimental apparatuses with which they are concerned. In fact, since no physical system is perfectly isolated from the rest of the Universe,* in order to make sure we are not missing anything we should, to be precise, take the whole Universe as the only physical system worth investigating. Otherwise we run the danger of neglecting some influence, say from the Crab Nebula, that might be crucial. Then states of cause and effect subsystems would be characterized by an infinite number of parameters; we would have to take

---

* Even if a physical system were isolated from the rest of the Universe with respect to all the known forces and interactions, there would still remain that influence which, according to the Mach principle (Ernst Mach, Austrian physicist, psychologist, and philosopher, 1838–1916), is the origin of inertia. It would indeed be a great day if we were to succeed in abolishing inertia. In addition, the effect of quantum entanglement is also an anti-isolatory factor. Quasi-isolated systems might be correlated with their surroundings via quantum correlations.

into consideration the state of every object, field, etc. in the Universe, and at all times too, since they might possibly influence the states of subsystems.

That holistic point of view, with its emphasis on the uniqueness of situations, is unconducive to science. (See the discussion in Section 9.2.) In its light everything appears so awesomely complicated that there seems to be no hope of understanding (in the sense of science) anything. Reproducibility and predictability seem utterly beyond reach. Science begins to be possible only when some order is discerned within the confusion, when similarities are remarked upon in spite of the differences. To do science one must, by decision, guess, or blissful ignorance, make assumptions about which influences are more or less important and on those assumptions "slice" the world into quasi-isolated components. One must then investigate those relatively simple physical systems as if they were really isolated. The simplest systems must be investigated first. Then more complicated systems can be attacked by considering them as syntheses of simpler systems in interaction. That very rough picture gives an idea of the attitude necessary for science to be possible.

Let me add that we have here, it seems to me, the main reason why science, and especially its predictability aspect, developed in the West and not in the East. While the dominant Eastern philosophies emphasized the oneness and wholeness of everything, making science extremely unlikely, if not altogether impossible, the Western *Weltansicht* encouraged analysis, making science most likely, if not inevitable. The different conceptions of the position of *Homo sapiens* vis-à-vis the rest of nature, tying in with religious considerations, were part of the whole scene. But let us return to our business.

We are thus led to the recognition that all scientific laws are doomed in principle to a sort of imprecision. The reasoning is that, if we took everything into consideration, we would be dealing solely with unique situations, reproducibility and predictability would be impossible, no order could be discerned within the confusion, and science would be inconceivable. It is only by ignoring certain aspects of physical reality that we can discover similarities, obtain reproducibility and predictability, discern order, and find laws.

That essential ignoring is usually taken to be of two kinds. There is *ignoring in practice*, where we admit there exists or might exist influence, but we assume it to be sufficiently weak that it is negligible compared with the effects under consideration. And there is *ignoring in principle*, where we assume there is no influence at all. An example of ignoring in practice is the influence of the star Sirius on our laboratory experiments. We know there are gravitational and electromagnetic influences, but we ignore them, because we know they are negligible. We do not know whether there are other influences, but we assume that, if there are, they too are negligible. An example of ignoring in principle is the influence of "absolute" position on our experiments; we assume there is none. We assume the laws of nature are the same everywhere in the universe.

From the empirical point of view it is impossible to distinguish between precise lack of influence and sufficiently weak influence. Experimental investigations can only supply upper bounds for the strength of the effect. Indeed, it is possible that *in principle* there is no precise lack of influence and that all ignoring is ignoring in practice. (Such matters and others are discussed in [M49].) *In practice*, however, it is convenient to think in terms of the two kinds of ignoring.

Return now to a law of behavior for a physical system in terms of antianalog computer and translation rules, as described previously. We have just convinced ourselves that the translation of states of a cause subsystem is essentially "weak" in the sense that certain aspects of the states are ignored in translation. Thus more than one state are translated into the same input to the antianalog computer. And similarly, the output does not contain sufficient information for unique determination of a state of the effect, so that more than one state of the effect is consistent with the translation of an output. That defines equivalent relations in the state spaces of the cause and effect subsystems, according to which states of the cause are equivalent if and only if they translate into the same input, and states of the effect are equivalent if and only if they are consistent with the translation of the same output.

For example, if a law is consistent with the special theory of relativity, all states differing only by spatial or temporal displacements, rotations, or Lorentz transformations are equivalent. States that differ only in that way translate into the same input or are translated from the same output for a relativistic law. In other words, a relativistic law may take interest in any property of a state that is not a position, instant, orientation, or velocity connected with that state.

Or, if a law for system of particles does not distinguish individual particles, all states differing only by permutation of particles are equivalent.

By that definition of equivalence in the state space of a cause subsystem, all states belonging to the same equivalence subspace translate into the same input. That input uniquely determines an output according to the law. And all states of the effect subsystem that are correlated with those states of the cause by the causal relation are translated from that output and are therefore equivalent to each other by definition. It is possible that also states of the effect that are correlated with states of the cause belonging to a different equivalence subspace, i.e., that translate into a different input, are equivalent to those, since different states of a cause may be causally related to the same state of an effect (but not vice versa) and since different inputs of a theory may determine the same output (while the same input always determines the same output). Figure 5.5 should help make that clear. Note in the example of the diagram that the state space of the cause decomposes into three equivalence subspaces, which translate into three different inputs. The antianalog computer converts two of those inputs into the same output and the third input into a different output. The two outputs translate into states of the effect, which decompose accordingly into two equivalence subspaces.

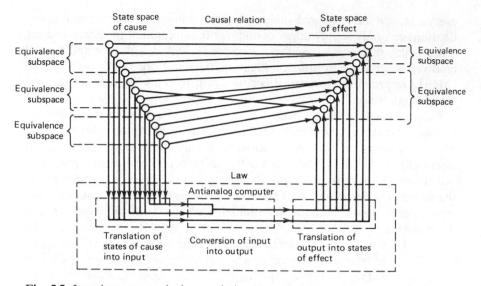

Fig. 5.5. Law imposes equivalence relations on state spaces of cause and effect in physical system, leading to the equivalence principle.

What we have just proved is *the equivalence principle*:

Equivalent causes—equivalent effects.

That formulation is relatively imprecise to avoid putting off those who have not yet studied its significance. The intention of the word "causes" is, of course, states of a cause, and the word "effects" similarly stands for states of an effect of the cause. The precise formulation is, therefore:

Equivalent states of a cause—
equivalent states of its effect.

And if we desire to put even more emphasis on the causal relation, we might insert an arrow:

Equivalent states of a cause →
equivalent states of its effect.

We should mention that the converse of the equivalence principles does not hold as a principle, since equivalent states of an effect are not necessarily causally correlated with equivalent states of a cause, as we saw.

We derived the equivalence principle as a direct consequence of the existence of science. Thus the existence of science requires the existence of the equivalence principle, which is therefore a necessary condition for the existence of science. No equivalence principle (and thus no reproducibility) → no science!

The equivalence principle will be empty if and only if the equivalence relation a law imposes on states of a cause is the trivial one, that every state is equivalent only to itself. Such a law does not allow reproducibility, so it is not suitable for science. Thus it is the "weaknesses," or "impotencies," of scientific laws, their essential inability to distinguish among the members of certain sets of states, that make the equivalence principle nontrivial.

Let us summarize our line of reasoning that led to the equivalence principle. We started with the fact that experimentally reproducible phenomena exist in nature and that causal relations are found in nature. We assumed the validity of predictability, i.e., we assumed that the human intellect is capable of exploiting reproducibility to attain sufficient understanding of natural causal relations to enable prediction of as yet unobserved phenomena. The conceptual tool for the expression of causal relations in physical systems is the law. Thus our assumption is that the human intellect is capable of inventing laws that describe the behavior of physical systems. That assumption seems to be well justified. A law, by its essential "impotency," imposes an equivalence relation in a cause state space and in an effect state space: States of a cause are equivalent if and only if they translate into the same input, and states of an effect are equivalent if and only if they are translated from the same output. From that and from the character of any causal relation we obtained the equivalence principle:

Equivalent states of a cause →
equivalent states of its effect.

Or in less precise language:

Equivalent causes—equivalent effects.

PROBLEMS

1. The cause is hearing a piece of music, and the effect is the amount of pleasure (or displeasure) experienced by the listener. The equivalence relation of the cause is that pieces of music by the same composer are considered equivalent. Describe the causal relation that makes the equivalence principle hold.

2. The charges, positions, and velocities of source charges and test charges are the cause, and the accelerations of the test charges are the effect. Could states of the cause that are identical except that all velocities are reversed and all charges have opposite sign be equivalent? If so, what is the narrowest allowed equivalence relation of the effect?

3. Let the cause be the set of all forces acting on a body, and let the effect be the linear and angular accelerations of the body. Describe the most comprehensive equivalence relation of the cause for each of the following equivalence relations of the effect: (a) States with the same direction of linear acceleration are equivalent; (b) states with the same magnitude of linear acceleration are equivalent; (c) states with equal linear accelerations are equivalent; and (d) states whose linear and angular accelerations are respectively equal are equivalent.

4. The system consists of crossed, uniform, electric and magnetic fields and a non-relativistic charged particle moving in the fields, where the particle enters the fields

moving perpendicularly to both. States of the particle are characterized by its mass, charge, and initial speed. For fixed electric field strength and given particle state the magnetic field strength can always be set so that the particle will pass through the crossed fields with no deflection. Let the state of the particle be the state of the cause and the magnetic field strength for zero deflection the effect. What is the most comprehensive equivalence relation of the cause?

5. The system is a nonrelativistic charged particle moving in a uniform magnetic field perpendicularly to the field. Such a particle moves in a circular orbit. States of the cause are characterized by the mass, charge, and speed of the particle, and strength of the magnetic field. The effect is the particle's orbit. What is the most comprehensive equivalence relation of the cause, if orbits of the same radius are considered equivalent?

6. It might happen that the phenomena of interest have to do with systems whose components are rather strongly interacting in general, so that their analysis into quasi-isolated, relatively simple subsystems is not at all straightforward or perhaps even impossible altogether. That might be the case for social, economic, and physchological phenomena, for example. Discuss reproducibility, predictability, order, disorder, and the possibility of finding scientific laws in such cases. Consider parapsychological phenomena in that light.

5.3. The Symmetry Principle

Although the equivalence principle is fundamental to the application of symmetry in science, in practice it is another principle, called *the symmetry principle* and derived directly from the equivalence principle, that is usually used:

> *The symmetry group of the cause is a subgroup*
> *of the symmetry group of the effect.*

Or less precisely:

> *The effect is at least as symmetric as the cause.*

See [A17].

To prove the symmetry principle we first define the terms used in it. Consider again a physical system and a law of its behavior. Any state u of the system implies a state for each of its subsystems and implies especially state u_c for a given cause subsystem and state u_e for a given effect subsystem. The law imposes an equivalence relation for the state space of the cause subsystem (giving the "equivalent causes" in the equivalence principle) and an equivalence relation for the state space of the effect subsystem (giving the "equivalent effects" in the equivalence principle), as discussed in the preceding section.

Define cause-equivalence, denoted $\overset{c}{\equiv}$, for the state space of the whole system as follows: Two states of the whole system are cause-equivalent if and only if the states of the cause subsystem implied by them are equivalent.

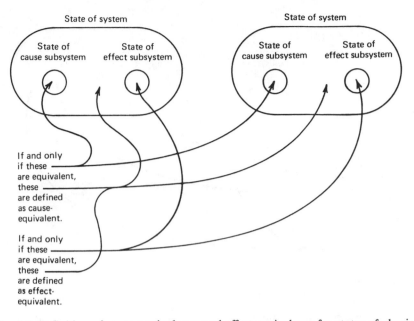

Fig. 5.6. Definition of cause-equivalence and effect-equivalence for states of physical system.

Symbolically,

$$u_c \equiv v_c \quad \Leftrightarrow \quad u \stackrel{c}{\equiv} v$$

for states u and v of the whole system. Similarly define effect-equivalence, denoted $\stackrel{e}{\equiv}$, for states of the whole system: Two states of the whole system are effect-equivalent if and only if the states of the effect subsystem implied by them are equivalent.

$$u_e \equiv v_e \quad \Leftrightarrow \quad u \stackrel{e}{\equiv} v$$

for states u and v of the whole system. See Fig. 5.6.

The equivalence principle states that if the states of the cause subsystem implied by states of the whole system are equivalent, then the states of the effect subsystem implied by those same states of the whole system are also equivalent. Symbolically,

$$u_c \equiv v_c \quad \Rightarrow \quad u_e \equiv v_e$$

for states u and v of the whole system. From our definitions we conclude that cause-equivalence implies effect-equivalence for the state space of the whole system. Or, symbolically,

$$u \stackrel{c}{\equiv} v \quad \Rightarrow \quad u \stackrel{e}{\equiv} v$$

for states u and v of the whole system. See Fig. 5.7.

We now define the symmetry group of the cause as the symmetry group of

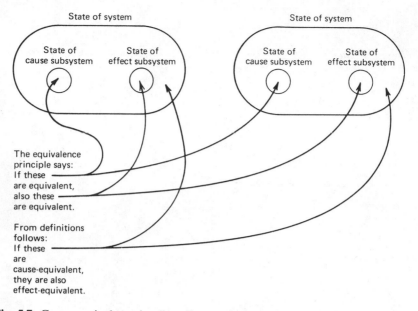

Fig. 5.7. Cause-equivalence implies effect-equivalence for states of physical system.

the whole system for cause-equivalence in its state space. Similarly, the symmetry group of the effect is defined as the symmetry group of the whole system for effect-equivalence. Since cause-equivalence implies effect-equivalence, it follows that every element of the symmetry group of the cause must necessarily also be an element of the symmetry group of the effect. There might, of course, be symmetry transformations of the effect that are not also symmetry transformations of the cause. That proves the symmetry principle, that the symmetry group of the cause is a subgroup of the symmetry group of the effect. And referring to our discussion of quantification of symmetry in Section 4.6, we obtain the alternative, less precise formulation of the symmetry principle, that the symmetry of the effect is at least that of the cause.

5.4. Minimalistic Use of the Symmetry Principle

The symmetry principle can be used to set a lower bound on the symmetry of the effect or to set an upper bound on the symmetry of the cause. The former use, which we call the minimalistic use, is characteristic of most practical, technological, and textbook problems, where the law and cause are known for a given system and it is desired to find some or all of the effect. If the cause is known, its symmetry can be worked out. The symmetry principle then states that the effect also possesses that symmetry (and possibly more). The knowledge of the minimal symmetry of the effect is often sufficient for solving the problem fully or partially or at least for simplifying it to some degree. The minimalistic use of the symmetry principle is exploited in science in widely varying degrees of sophistication. Here we offer some examples of the more simple-minded applications, in order to keep the symmetry considerations in the foreground.

In our first example we find the direction of acceleration of a resting test charge due to a spherically symmetric charge distribution. What we are doing, of course, is finding the direction of the electric field of a spherically symmetric charge distribution, but we prefer to formulate the problem in spatiotemporal terms, in terms of acceleration, rather than in terms of the electric field, since the transformation properties of the electromagnetic field are not trivial, and we prefer not getting involved with them here. Note that we also avoid formulating the problem in terms of force, since, although no harm would be done, we still feel we are standing on firmer ground when we assign transformation properties to acceleration rather than to force. Also note that this example is equally applicable to any central force field, as no use is made of the Coulomb force law (Charles Augustin de Coulomb, French physicist, 1736–1806) specifically.

The cause in this example consists of the charge distribution and the test charge. The effect is the acceleration of the text charge. Refer to Fig. 5.8. The symmetry group of the cause is the group of all rotations about the axis passing through the test charge and the center of the charge distribution and reflections through all planes that contain that axis. By the symmetry principle the effect, the acceleration, must also have that symmetry at least. Since any nonradial component the acceleration might have would not be carried into itself by all the symmetry transformations of the cause, the acceleration must therefore be radial. See [A54].

In the next example we find the direction of acceleration of a test charge moving parallel to a straight, infinitely long, current-carrying wire. Although the force on the test charge is magnetic, that fact will not enter our discussion. And we purposely formulated the problem in terms of the acceleration of a test particle, rather than in terms of the magnetic field around the wire, in order to avoid the complicating side issue of the transformation properties of the magnetic field under reflection, a common source of trouble to the unwary.

The cause consists of the current and the moving test charge. The effect is

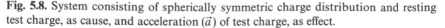

Fig. 5.8. System consisting of spherically symmetric charge distribution and resting test charge, as cause, and acceleration ($\vec{a}$) of test charge, as effect.

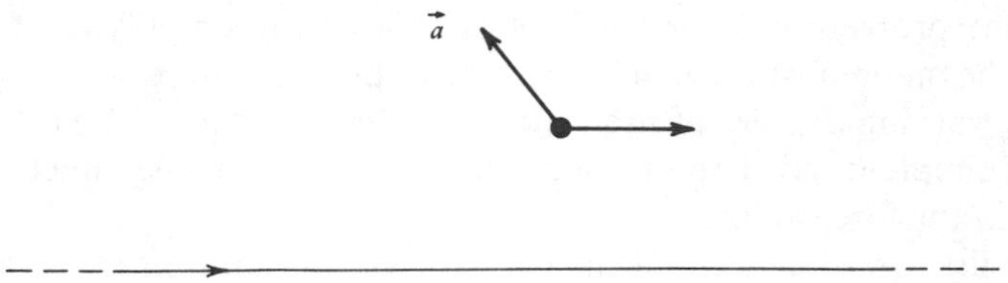

Fig. 5.9. System consisting of infinite straight current and test charge moving parallel to it, as cause, and acceleration ($\vec{a}$) of test charge, as effect.

the acceleration of the test charge. Refer to Fig. 5.9. (Although in the figure the current was given the same sense as the velocity of the test charge, that point is immaterial, and the opposite sense would serve as well.) The symmetry group of the cause contains the transformation of reflection through the plane containing the wire and the test charge (the plane of the paper in the figure). Since by the symmetry principle that must also be a symmetry transformation of the effect, the acceleration can have no component perpendicular to that plane.

The symmetry group of the cause also contains the more complicated transformation consisting of the consecutive application of temporal inversion and spatial reflection through the plane perpendicular to the current and containing the text charge (at any instant). The temporal-inversion transformation by itself reverses the senses of all velocities, in our example both that of the test charge and that of the moving charges forming the current. Since both the current and the velocity of the test charge are perpendiclar to the reflection plane of the second part of the transformation, their senses are reversed again by the reflection transformation. Thus the combined transformation indeed leaves the cause intact. See Fig. 5.10. But how does it affect the effect, the acceleration, of which it must be a symmetry transformation? Temporal inversion by itself does not change acceleration, since the dimension of acceleration is even-powered in time, (length) $\times$ (time)$^{-2}$. And the reflection reverses the component of acceleration perpendicular to the reflection plane. See Fig. 5.10. So symmetry of the effect further limits the possible directions of the acceleration and leaves it perpendicular to the velocity of the test particle and pointing either toward or away from the wire.

In our next example we prove that the orbit of a planet about its sun lies completely in a plane and that this plane passes through the center of the sun. Our only assumption is that the sun and planet are each spherically symmetric. We assume nothing about the nature of the force between the two. (The reasoning will be valid also for a spherically symmetric body in any central force field.)

Consider the situation at any instant. The planet has a certain position and a certain instantaneous velocity relative to the sun. Of all the planes containing the line connecting the centers of the planet and the sun only one is parallel to the direction of the planet's instantaneous velocity. We call

Fig. 5.10. Symmetry transformation of cause for system of Fig. 5.9.

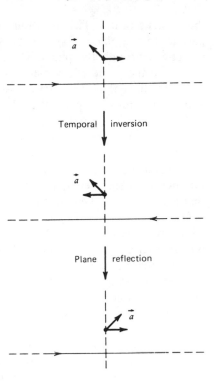

that the plane of symmetry. See Fig. 5.11. The cause consists of the sun and the moving planet. The effect is the planet's acceleration.

The cause has reflection symmetry with respect to the plane of symmetry: The sun and planet are each reflection symmetric with respect to any plane containing their centers, since they are spherically symmetric, and the plane of symmetry contains both centers. The position of the planet is not changed by reflection through the plane of symmetry, again since the plane contains

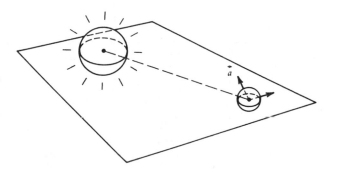

Fig. 5.11. System consisting of sun and moving planet, as cause, and acceleration ($\vec{a}$) of planet, as effect. Plane of symmetry is indicated.

the planet's center. The direction of the planet's instantaneous velocity is parallel to the plane of symmetry, so that velocity is also invariant under reflection through the plane. Thus the cause is reflection symmetric with respect to the plane of symmetry. The effect, the acceleration of the planet, must then have that symmetry at least. Therefore, the direction of the planet's acceleration must be parallel to the plane of symmetry. (If the acceleration had a component perpendicular to the plane, it and its reflection image would not coincide.) So the planet's velocity, which is parallel to the plane of symmetry, undergoes a change of velocity parallel to the plane of symmetry and thus remains parallel to that plane. In that way we see that the planes of symmetry of the system at all instants are in fact one and the same plane and that the motion of the planet is confined to that plane.

Our next example of application of the symmetry principle concerns electric currents. Consider the DC circuit of Fig. 5.12, certainly good for nothing but an exercise. The six emf (voltage) sources, the 11 resistors, and their connections are the cause. The effect consists of the resulting currents in the various branches of the circuit and the potential differences between all pairs of points of the circuit. By Kirchhoff's first law (Gustav Robert Kirchhoff, German physicist, 1824–1887) we arbitrarily designate seven of the currents $i_1, i_2, i_3, i_4, i_5, i_6, i_7$ and express the other six currents in terms of those as in the figure. Kirchhoff's first and second laws then give us a set of seven simultaneous linear equations for the seven unknown currents, which we do not show here. Those equations can be solved and the solutions used to calculate the other six currents. With all the currents known the

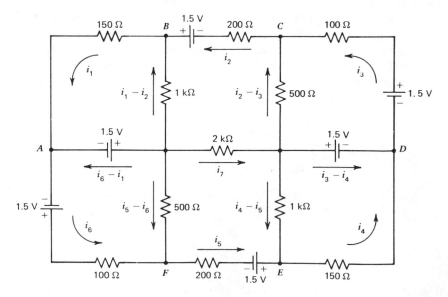

Fig. 5.12. DC circuit.

potential difference between any pair of points can be calculated. For example, referring to the figure, we might want to know the potential difference between any pair of the points marked A, B, C, D, E, F, such as V_{AB}, V_{AD}, V_{EF}.

A glance at Fig. 5.12 reveals that the cause has symmetry, twofold rotation symmetry about the axis through the 2-kΩ resistor and perpendicular to the plane of the paper. So the symmetry principle can be invoked, and the effect, the currents and potential differences, must be at least as symmetric as that. To help us see what that gives we rotate the system by 180°, as in Fig. 5.13. Now compare the rotated system of Fig. 5.13 with the unrotated system of Fig. 5.12. Symmetry of the effect gives us for the currents $i_1 = i_4$, $i_2 = i_5$, $i_3 = i_6$, $i_7 = -i_7$, so that $i_7 = 0$. For the potential differences symmetry of the effect gives $V_{AD} = V_{DA}$, $V_{BE} = V_{EB}$, $V_{CF} = V_{FC}$. And since by the nature of potential differences $V_{DA} = -V_{AD}$, etc., we have $V_{AD} = V_{BE} = V_{CF} = 0$. Symmetry of the effect also gives us relations among the potential differences, such as $V_{AF} = V_{DC}$, $V_{CE} = V_{FB}$, etc.

Thus the symmetry principle gives us more than half the solution to the problem in this example. Instead of solving seven simultaneous equations for seven unknowns, we need to solve only three equations for three currents. Three of the desired potential differences are now known, and only half of the others need to be calculated.

Our next example is again an electrical one. In this example the system has a higher degree of symmetry than in the preceding one, and the symmetry principle makes the complete solution quite simple, although the explaining

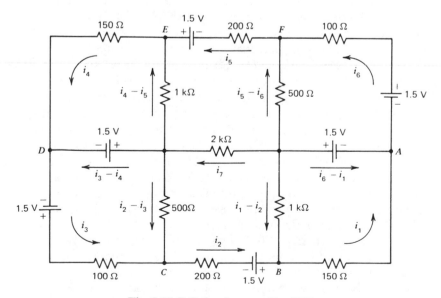

Fig. 5.13. DC circuit rotated by 180°.

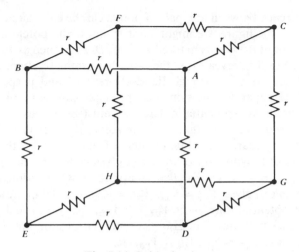

Fig. 5.14. Cube of resistors.

of it is a bit drawn out. We solve the well-known problem of finding the resistance of a network of 12 equal resistors connected so that each resistor lies along one edge of a cube, where the resistance is measured between diagonally opposite vertices of the cube, such as between vertices A and H in Fig. 5.14. Let r denote the resistance of each resistor. We imagine a potential difference V applied between vertices A and H. As a result current i enters the network at A, branches through the various resistors, and leaves the network at H. The resistance of the network from A to H is then $R = V/i$ by Ohm's law (Georg Simon Ohm, German physicist, 1787–1854). The cause is the network and the applied potential difference, while the currents in the resistors and the corresponding potential drops comprise the effect.

The cause in the present example does not possess the full symmetry of the cube, in spite of all the resistors being equal, since vertices A and H are distinguished from the other vertices and from each other, as the current enters the network at A and leaves at H. So the symmetry transformations of the cause consist of only those symmetry transformations of the cube that do not affect vertices A and H: rotations by 120° and 240° about the diagonal AH (i.e., diagonal AH is an axis of threefold rotation symmetry) and reflections through each of the three planes $ABHG$, $ACHE$, and $ADHF$ as in Fig. 5.15. Then by the symmetry principle the effect must also have that symmetry. We can use that fact to find the current in each resistor of the network.

The current i entering the network at vertex A splits among three branches and flows to vertices B, C, and D. Due to the threefold rotation symmetry the current divides equally, so that current $\frac{1}{3}i$ flows in each branch AB, AC, and AD as in the diagram of Fig. 5.16. The current $\frac{1}{3}i$ entering vertex B then divides again between two branches and flows to vertices E and F. It divides equally between the two branches because of the reflection symmetry with

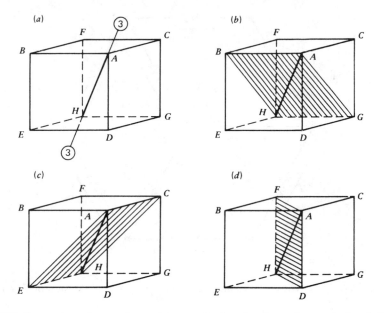

Fig. 5.15. Symmetry of cause for resistance of cube of resistors between A and H. (a) Axis of threefold rotation symmetry. (b)–(d) Planes of reflection symmetry.

respect to plane $ABHG$. Thus current $\frac{1}{2}(\frac{1}{3}i) = \frac{1}{6}i$ flows in each branch BE and BF. Similar reasoning also gives current $\frac{1}{6}i$ in each of the branches CF, CG, DG, and DE as in Fig. 5.17.

After that stage the symmetry takes care of itself. Current $\frac{1}{6}i$ enters vertex E from each of vertices B and D, producing current $2 \times \frac{1}{6}i = \frac{1}{3}i$ leaving E and flowing to H. Similarly, current $\frac{1}{3}i$ enters H from each of vertices F and G.

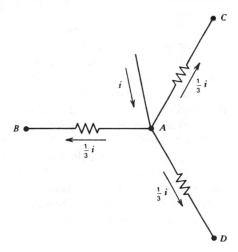

Fig. 5.16. Rotation symmetry requires equal first division of current entering network.

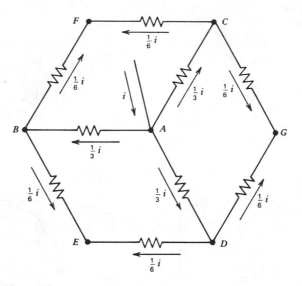

Fig. 5.17. Plane-reflection symmetry requires equal second division of current.

The three currents $\frac{1}{3}i$ entering vertex H join to give current $3 \times \frac{1}{3}i = i$ leaving the network. That is illustrated in Fig. 5.18.

Now, the potential difference V between A and H equals the sum of potential drops between those two vertices, where the sum may be calculated over any continuous path connecting A and H. Let us take path $ABEH$. By Ohm's law the potential drop on a resistor equals the product of the current in the resistor and its resistance. So, referring to Fig. 5.18, the potential drop from A to B is $\frac{1}{3}ir$. From B to E the potential drop is $\frac{1}{6}ir$, and from E to H it is $\frac{1}{3}ir$. Adding those together, we obtain $V = \frac{5}{6}ir$. Thus resistance of the network between A and H is

$$R = V/i = \frac{5}{6}ir/i = \frac{5}{6}r.$$

That is the solution of the problem.

In the last two examples note that it is not the actual geometry of the circuits that is important but rather their electric structure. Thus it is not the geometric symmetry of the circuits that really interests us but rather the permutation symmetry of their components. It is true that we expressed that permutation symmetry in geometric terms in order to keep our considerations as familiar as possible; however, the former circuit does not have to be laid out as nicely as in Fig. 5.12 nor does the latter network actually have to lie along the edges of a cube for our symmetry arguments to hold. Thus it is really permutation symmetry that we are using, based on the functional, rather than geometric, equivalence of parts of the circuit. That point is important to keep in mind in the application of symmetry to physical systems: Functional equivalence of parts of a system can be a source of permutation symmetry.

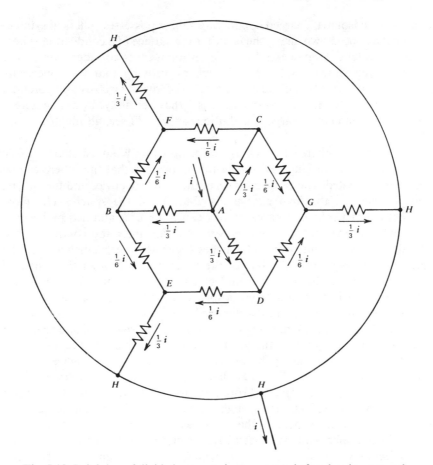

Fig. 5.18. Rejoining of divided currents in two stages before leaving network.

Another point to note in the last two examples is that each problem has a unique solution. So naturally the solution was chosen to be the effect in each case, and application of the symmetry principle was straightforward. Uniqueness of solution is characteristic of passive electric circuits. That is not always the case, however, when circuits contain active components (transistors, vacuum tubes, etc.). Then more than one solution might be possible. What is the effect in that case and how is the symmetry principle applied? Read on.

Our next example of application of the symmetry principle is of somewhat different character. It is often assumed that the electric charge distribution on the surface of an isolated charged conducting sphere is homogeneous, i.e., that the surface charge density in such a case possesses spherical symmetry. And if any explanation is offered at all, it is invariably simply a reference to "symmetry considerations." Now it is indeed true that the surface charge density of an isolated charged conducting sphere possesses spherical symmetry. That can be shown, using the fact that the surface of a conducting sphere is an equipotential surface and relying on uniqueness considerations. That has not been shown, as far as I know, by proving that the spherically symmetric configuration is the configuration of lowest electrostatic potential energy. As for "symmetry considerations," they certainly do not require a spherically symmetric charge distribution, as we will see, although they do allow it.

How does the sphere get its charge to begin with? We must charge it. We then remove the charging device far enough away so that the sphere can be considered isolated. The cause consists of the charging device and the sphere, and the effect is the final charge distribution on the sphere. The most symmetric practically attainable cause I can think of is attained by keeping the charge source very far from the sphere at all times and connecting the two through a straight wire that touches the sphere perpendicularly to the sphere's surface, so that the line of the wire passes through the center of the sphere. When charging is completed, the wire is withdrawn from the sphere along its line. (Alternatively, a spherical charge source could be put in contact with the sphere and then removed along the line containing the two centers.) Thus the most symmetry we can give the cause is axial and reflection symmetry, symmetry under the group of all rotations about the line of the wire (or the line containing the centers) and reflections through all planes containing that line. See Fig. 5.19. That is not spherical symmetry, which is symmetry under the group of all rotations about all axes through the center of the sphere and reflections through all planes containing the center. The latter group obviously includes the former as a proper subgroup. Indeed, the sphere itself is spherically symmetric. However, the *total* cause is only axially and reflection symmetric. So the only symmetry we can demand of the effect, the charge distribution, is also axial and reflection symmetry, although spherical symmetry is not excluded by the symmetry principle.

If we do not know or prefer to ignore the way the sphere is charged and wish to consider the sphere itself as the cause, the effect will certainly not be the charge distribution, which might depend on the way the sphere is charged, but the set of all allowed charge distributions. Since the cause in this case is spherically symmetric, so is the effect, by the symmetry principle. Thus, even though no single allowed charge distribution has to be spherically

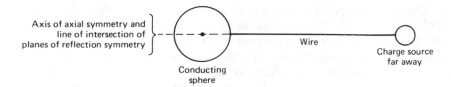

Fig. 5.19. Charging of conducting sphere.

symmetric, given any one allowed charge distribution, all those obtained from it by any rotation about any axis through the center of the sphere or reflection through any plane containing the center are also allowed charge distributions. See [A54.]

That brings us to the following warning. If the state of what is chosen as the cause in a problem is not sufficient to determine a unique solution to the problem, then the effect cannot be the solution, but is rather the family of all solutions consistent with a state of the cause. As another example of that, given as the cause an equation possessing some symmetry, the effect is not any solution of the equation, and indeed no solution need possess the full symmetry of the equation. The effect is the family of all solutions of the equation, and that is what must have the symmetry of the equation. Given any one solution, the application to it of any symmetry transformation of the equation produces another solution (or even possibly the same solution). If, however, additional conditions are involved, such as boundary or initial conditions or constraints, they may be included in the cause and will usually reduce its symmetry.

To give a more specific example, consider the algebraic equation

$$x^2 - 4 = 0$$

or any other polynomial equation involving only either even or odd powers of the unknown variable. Every such equation is symmetric under the transformation

$$x \to -x.$$

(The odd polynomials change sign under that transformation, but the equations are the same.) If each root of such an equation had the same symmetry, the only allowed root would be $x = 0$. However, it is the set of all roots, and not each root, of the equation that must possess the symmetry of the equation, with the result that for equations of the type we are discussing all nonzero roots must come in positive-negative pairs. For instance, the roots of the equation just presented are $x = \pm 2$.

Or consider the harmonic-oscillator differential equation

$$\frac{d^2 y}{dt^2} + a^2 y = 0, \qquad a = \text{const.}$$

It possesses symmetry under each of the following transformations:

$$\text{temporal displacement} \quad t \to t + b$$

for arbitrary constant b,

$$\text{temporal inversion} \quad t \to -t,$$

$$\text{spatial dilation} \quad y \to cy,$$

for arbitrary, positive constant c, and

$$\text{spatial inversion} \quad y \to -y.$$

Since no initial conditions are specified, the family of solutions of this equation must have the same symmetry as the equation. Thus, if

$$y = f(t)$$

is a solution, so must also

$$y = f(t + b),$$
$$y = f(-t),$$
$$y = cf(t),$$
$$y = -f(t),$$

be solutions for arbitrary allowed constant b and c. For the oscillator equation the family of all solutions can be written in the form

$$y = A \sin(at + B) = C \sin at + D \cos at$$

for all constant A and B or C and D. Under the temporal-displacement transformation the general solution becomes

$$y = A \sin(at + ab + B)$$
$$= A \sin(at + B'),$$
$$B' = ab + B.$$

Since B' ranges over the same values as B, we have recovered the same family. Under the temporal-inversion transformation the general solution becomes

$$y = A \sin(-at + B)$$
$$= -A \sin(at - B)$$
$$= A' \sin(at + B'),$$
$$A' = -A,$$
$$B' = -B,$$

and again we find that the family is symmetric. Under the spatial-dilation transformation the general solution becomes

$$y = cA \sin(at + B)$$
$$= A' \sin(at + B),$$
$$A' = cA;$$

symmetry again. And under the spatial-inversion transformation the general solution becomes

$$y = -A \sin(at + B)$$
$$= A' \sin(at + B),$$
$$A' = -A,$$

and is symmetric. If initial conditions are imposed by specifying the values of the solution and its derivative for some value of t, the particular solution will in general not possess any of the above symmetries, although in certain cases it might retain one or more of them. See [M58].

Note that each particular solution of the oscillator equation has the property of periodicity, i.e., it is symmetric under a certain minimal temporal displacement and integer multiples of it,

$$y\left(t + \frac{2\pi}{a}n\right) = y(t)$$

for

$$n = 0, \pm1, \pm2, \ldots.$$

(That symmetry group is a subgroup of the temporal-displacement group.) That symmetry does not follow from the symmetry of the equation but can be discovered only by actually solving the equation. In comparison, the equation

$$\frac{d^2y}{dt^2} - a^2y = 0,$$

similar to the oscillator equation, possesses the same four symmetries that we found for the oscillator equation but does not have periodic solutions. In fact, its general solution can be written

$$y = C \sinh at + D \cosh at,$$

and the hyperbolic functions are not periodic. This family of solutions must and does possess the four symmetries of its equation.

PROBLEMS

1. Extend the discussion of the spherically-symmetric–charge-distribution example, the first example of this section, to show that for given charge distribution the only geometric variable that the magnitude of the test charge's acceleration can depend on is the distance between the test charge and the center of the distribution.

2. For the example of a current-carrying wire and moving test charge extend the argument to show that the only geometric variable on which the magnitude of the test charge's acceleration can depend is the distance between the test charge and the wire.

3. For a uniformly charged infinite plane prove by the symmetry or equivalence principle that: (a) the direction of the electric field produced by such a charge distribution is perpendicular to the plane; (b) the only geometric variable that the field strength can depend on is the distance from the plane; and (c) at equal distances on either side of the plane the field strengths have equal magnitudes and opposite senses.

4. In the sun-and-planet example of this section the plane of symmetry is not uniquely defined when the planet is at rest or moving directly toward or away

from the sun. Use the symmetry principle to prove that in those cases the orbit of the planet is on a straight line through the center of the sun.

5. Using the symmetry principle, show for the cube of resistors discussed in this section that vertices B, C, and D are equipotential points and so are vertices E, F, and G. Prove that also by electrical considerations in the diagram of Fig. 5.18.

6. Use the symmetry principle to find the resistance of a network of 12 equal resistors connected to form the edges of an octahedron, where the resistance is measured between a pair of opposite vertices.

7. According to Hermann Weyl (German–American mathematician, 1885–1955) [S31], Ernst Mach "tells of the intellectual shock he received when he learned as a boy that a magnetic needle is deflected in a certain sense, to the left or to the right, if suspended parallel to a wire through which an electric current is sent in a definite direction." See Fig. 5.20. Explain why that phenomenon seems to violate the symmetry principle, to young Mach's shock. Explain why it really does not.

8. Find a useful or interesting nontrivial problem that can be completely solved or at least greatly simplified by the symmetry principle.

9. Find the symmetries of each of the following equations and, solving the equation, show that the set of roots of each equation possesses the same symmetries:
   (a) $x^2 + 4 = 0$;
   (b) $x^5 - x^3 - 6x = 0$; and
   (c) $\cos x = \sqrt{1 - \sin^4 x}$.

10. Find the symmetries of each of the following differential equations and show, by solving it, that the general solution of each equation possesses the same symmetries. Then impose the initial condition $y(0) = 1$ on each solution and check which of the symmetries are lost and which are retained.
   (a) $\dfrac{d^2 y}{dt^2} = a$, $a = \text{const}$;

   (b) $\dfrac{dy}{dt} = ay$, $a = \text{const}$; and

   (c) $y\dfrac{dy}{dt} = a$, $a = \text{const}$.

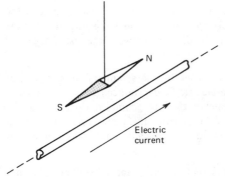

Fig. 5.20. Mach's dilemma.

11. Use the symmetry principle to show that the complex roots of a polynomial equation with real coefficients must be in conjugate pairs. What is the symmetry transformation involved here?

5.5. Maximalistic Use of the Symmetry Principle

The problems of basic scientific research are usually opposite to those for which the symmetry principle is used minimalistically. In basic research it is the effect that is given, and we try to find the cause. The effect in such a problem is one or more experimental phenomena, and we attempt to discern order among them, find laws for them, and devise a theory explaining those phenomena as being brought about by some cause. A theory is considered to be better the more different phenomena it explains and the simpler the cause that is supposed to be producing them. See [M52]. Although simplicity is not a standardized concept and is largely a matter of taste, it is generally agreed that symmetry contributes greatly to simplicity. So in devising a theory for a given set of experimental phenomena, we usually assume as symmetric a cause as possible. And how symmetric can a cause be? Here the symmetry principle serves to set an upper bound on the symmetry of the cause; the cause can be no more symmetric than the effect. That is what we call the maximalistic use of the symmetry principle.

So we must first identify the symmetry of the phenomena we wish to explain. (That symmetry is often far from obvious.) We then construct our theory so that the cause will have just the same symmetry, if possible. If it is not possible to assign maximal symmetry to the cause, we must assume a less symmetric cause and include in the theory an explanation of why the effect is more symmetric than the cause. But we may never assume that the cause has a higher degree of symmetry than the phenomena being explained. That would violate the symmetry principle and thus also the equivalence principle.

What most often happens, though, is that the symmetry of a set of phenomena is only approximate. (We discuss approximate symmetry in Chapter 6; for the present let us make do with whatever idea about it, however vague, you might have or the discussion in Section 8.1, if you are on the "concept" track.) In such a case the first step toward a theory is to determine the ideal symmetry that is only approximated by the phenomena. That can be very difficult, if the symmetry is far from exact. Then to obtain as symmetric a cause as possible we try to construct a theory such that the cause will have a dominant part (in an appropriate sense), possessing the ideal symmetry of the effect, and another, symmetry-breaking part, which does not have that symmetry. In the (possibly hypothetical) limit of complete absence of symmetry breaking, the dominant part of the cause produces the ideal symmetry of the phenomena, while the symmetry-breaking part brings about the deviation from ideal symmetry. A complete theory will contain the symmetry-breaking mechanism within its framework. But sometimes the cause of the breaking must be left as a mystery to be cleared up when more experimental facts are known or a better theory can be found.

For an example of the maximalistic use of the symmetry principle we turn to nuclear physics. The basic problem of nuclear physics is the strong nuclear interaction, the force that binds protons and neutrons together to form nuclei. That nuclear force is not yet completely understood. And neither is the weak nuclear interaction, a force that also affects protons and neutrons, but at a strength much lower than that of the strong interaction. On the other hand, the electromagnetic interaction among protons and neutrons as well as the Pauli exclusion principle (Wolfgang Pauli, Swiss–Austrian physicist, 1900–1958), stating in the present context that no two protons or no two neutrons can be in the same quantum state, are very well understood. The effect consists of such phenomena as the various properties of all kinds of nuclei and the results of scattering experiments, in which protons and neutrons are made to collide. The effect is found to exhibit the following approximate symmetry. Two kinds of nuclei differing only in that one of the neutrons in one kind is replaced by a proton in the other often have certain similar properties (though their electric charges, for example, are clearly different); in scattering experiments similar results are obtained whether the interacting particles are two protons, two neutrons, or a proton and a neutron.

Thus nuclear phenomena are approximately symmetric under interchange of proton and neutron. That symmetry is called charge symmetry, since the major difference between the proton and the neutron is their different electric charges. The symmetry principle then suggests that we assume that the strong nuclear interaction is exactly charge symmetric, is completely blind to any difference between the proton and the neutron. The symmetry-breaking factors are assumed to be the electromagnetic interaction, which discriminates between proton (electrically charged) and neutron (electrically neutral), the weak nuclear interaction, which also discriminates between proton and neutron, and the Pauli principle, which discriminates between identical particles and different particles. That assumption has, in fact, proved to be very successful.

Another example of the maximalistic use of the symmetry principle is in the study of elementary particles and their interactions (of which nuclear physics is actually just a special case). Here the symmetries of the experimental phenomena are much more complicated and much more approximate (i.e., less nearly exact, except for certain exact symmetries) than in nuclear physics, and the theorists are searching for ways to take them into account in their theories.

PROBLEM

For those on the "concept" track: Reconsider nuclear charge symmetry in the hypothetical case when the electromagnetic interaction, the weak interaction, the Pauli exclusion principle, and any other possible factor that distinguishes between proton and neutron are absent. In light of the discussions in Sections 8.2 and 8.3, would this case be more symmetric than the real situation? Would the hypothetical case possess exact neutron–proton interchange symmetry, while in reality such symmetry is only approximate?

5.6. Quantum Systems

This section is intended for readers with a good understanding of quantum theory and its Hilbert-space formulation. If that description does not fit, you are advised to skip the section.

The application of symmetry in quantum systems is straightforward and successful. It is, in fact, one of the greater success stories of theoretical science. That is due to the special nature of quantum state spaces, specifically to their being linear vector spaces, which makes it possible to work with realizations of transformation and symmetry groups by groups of matrices acting on the components of the elements (vectors) of quantum state spaces. Such realizations are called linear representations, or representations. The detailed study of the application of symmetry in quantum systems, which is beyond the scope of this book, is well presented in a considerable number of more advanced texts.

It is worthwhile, however, to devote a brief discussion to causes and effects in quantum systems. Causes and effects, as defined previously, are subsystems of the systems under consideration, and subsystems of quantum systems are themselves quantum systems. Thus causes and effects in quantum systems are quantum systems with all the consequences thereof. For example, in a scattering experiment, if the sharply peaked momentum distribution in the incident beam is part of the cause, one may not expect that the positions of the individual particles in the beam might also be included in the cause, since the latter are in principle indeterminate in the given setup. As for the effect, neither particle trajectories nor scattering angles of individual particles can be part of it. The effect is, in fact, the wave function, especially the "scattered" wave function, from which probabilities and differential cross sections can be calculated.

If that sounds like the Rutherford experiment, you are quite right, and we should compare the classical and quantum theories of such scattering with regard to cause and effect in each kind of theory. In the classical approach the cause could in principle be the initial position, velocity, and orientation of each incident particle and the position and orientation of each target particle. The effect would then be the point where each scattered particle hits the fluorescent detection screen and causes it to flash. However, such a cause is not practically realizable (and is, of course, precluded by quantum principles), and we must make do instead with the statistical properties of the incident beam and the target material. The effect is then correspondingly reduced to statistical properties of the scattered particles. In nonrelativistic quantum mechanics, on the other hand, the cause is the incident wave function and the scattering potential, which indeed involve statistical properties of the incident and target particles but are different from the classical cause. For instance, the incident wave function has a nonclassical phase (over which we admittedly have no control, but which could be of importance in its relation to other phases in the system). The effect is the scattered wave function, which also contains a phase in addition to statistical information about the scattered particles.

The vacuum state of quantum theory is also due for some discussion. Invariably, even in the most fundamental theories, the properties of the vacuum state are part of the assumptions of the theory, rather than derived from other assumptions. Thus the vacuum state is part of the cause. That must be kept in mind, especially when the vacuum state is assumed to possess a lower degree of symmetry than the rest of the cause. It is the symmetry of the total cause, including the vacuum state, that, by the symmetry principle, must appear in the effect. Since the vacuum state is crucial in determining the properties of the physical states, as the effect, it should come as no surprise that the physical states are not as symmetric as the cause without the vacuum state, and the situation is in no way a violation of the symmetry principle.

I brought up that point because it has often been presented as an apparent violation of the symmetry principle. The symmetry principle has nothing to fear from such or other *apparent* violations. It has been contrived so that, as long as we are concerned with conventional science (reproducibility, predictability, laws, etc.), it is inviolable.

5.7. Summary of Chapter Five

In this chapter we proceeded with the development of the general symmetry formalism that was started in the preceding chapter and considered the theory and practice of the application of symmetry in science. We started by discussing in Section 5.1 the concept of causal relation in physical systems, whereby certain correlations exist between states of cause subsystems and states of effect subsystems, correlations resulting from the fact that states of subsystems are determined by the states of the whole system.

In Section 5.2 we looked into scientific laws as expressions of causal relations. We saw that such laws must ignore certain aspects of states of physical systems. That introduces equivalence relations in the state spaces of systems, from which follows the equivalence principle: Equivalent states of a cause → equivalent states of its effect. From the equivalence principle we derived in Section 5.3 the symmetry principle: The symmetry group of the cause is a subgroup of the symmetry group of the effect. Or less precisely: The effect is at least as symmetric as the cause.

The symmetry principle can be used to set a lower bound on the symmetry of the effect or to set an upper bound on the symmetry of the cause. The former, its minimalistic use, is characteristic of most practical and study problems, where the cause is given and the effect is to be found. It helps toward simplifying or even solving the problem. We saw examples of that in Section 5.4. But, as we saw in Section 5.5, the symmetry principle serves basic scientific research maximalistically, since there it is the effect that is known and the cause is to be found. We considered an example of that.

In Section 5.6 we briefly discussed causes and effects in quantum systems.

CHAPTER SIX

Approximate Symmetry and Spontaneous Symmetry Breaking

In this short chapter we briefly discuss the formalism of approximate symmetry, based on the concept of a metric for a state space of a system. We present the concepts of approximate symmetry transformation, approximate symmetry group, goodness of approximation, exact symmetry limit, broken symmetry, and symmetry-breaking factor. We then more substantially discuss spontaneous symmetry breaking, situations where the symmetry principle—that the effect is at least as symmetric as the cause—*seems* to be invalid. The essence of the matter is the stability of the effect under perturbations of the symmetry of the cause.

The previous chapters of this book are prerequisite for this chapter. No additional mathematical knowledge is needed. The scientific background of the preceding chapter should more than suffice. The paragraph concerned with quantum theory at the end of the chapter is intended, of course, for the initiated.

6.1. Approximate Symmetry

In Section 8.1 we present a conceptual definition of approximate symmetry. However with or without that preparation, in order to express approximate symmetry in the general symmetry formalism we are developing, we need the notion of *approximate symmetry transformation*, any transformation that maps every state to an image state that is "nearly" equivalent to the object state. And just what does "nearly" equivalent mean? For that we must soften the all-or-nothing character of the equivalence relation, upon which symmetry is based, in order to allow, in addition to equivalence, varying degrees of inequivalence. The way to do that is to define a *metric* for a state space of a system, a "distance" between every pair of states, such that zero "distance" indicates equivalence and positive "distances" represent degrees of inequivalence.

More precisely, a *metric* is a bivariable, nonnegative, real function on state space $d(\ ,\ )$ having the following properties for all states u, v, w:

1. *Null self-distance* $d(u, u) = 0,$
2. *Symmetry* $d(u, v) = d(v, u),$
3. *Triangle inequality* $d(u, w) \leq d(u, v) + d(v, w).$

A metric is a generalization of an equivalence relation and includes it as a special case, where

$$d(u, v) = 0$$

defines equivalence of states u and v, $u \equiv v$. That is indeed an equivalence relation, since the properties of (1) reflexivity, (2) symmetry, and (3) transitivity of an equivalence relation follow, respectively, from the properties of (1) null self-distance, (2) symmetry, and (3) triangle inequality of a metric. Thus for a state space equipped with a metric the symmetry group is defined as before.

For a state space equipped with metric $d(\ ,\ )$ some of the transformations might map all states to "nearby" states only. That is, for a given positive number ε there might be transformations T such that

$$d(u, T(u)) \leq \varepsilon$$

for all states u. Such a transformation is called an *approximate symmetry transformation* of the system. We see that whether a given transformation is an approximate symmetry transformation depends on the value taken for ε. The larger the value of ε, the more transformations are approximate symmetry transformations in general. The set of all invertible approximate symmetry transformations of a given system for any of its state spaces equipped with metric and for a given ε does not in general form a group (there might be trouble with closure), although it does, of course, include the corresponding symmetry group and might include other groups that in turn include the symmetry group.

Since the general theory of approximate symmetry is not very well developed. I do not think it worthwhile to go into many details. It suffices to state that it is possible to define *approximate symmetry groups* for state spaces equipped with metrics, and it is possible to define a measure of *goodness of approximation* for each approximate symmetry group. A system will in general have different approximate symmetry groups with different measures of goodness of approximation. But for the same state space and metric any approximate symmetry group will include as subgroups all approximate symmetry groups of better approximation, and, especially, every approximate symmetry group will include the corresponding symmetry group as a subgroup. Thus the approximate symmetry groups and exact symmetry group of a state space equipped with metric can be arranged in an inclusion chain, with the exact symmetry group at the included end and worsening approximation in the direction of more inclusive groups. An approximate symmetry group does not in general consist of all the invertible approximate symmetry transformations for some value of ε. And, perhaps somewhat strangely, for any given value of ε an approximate symmetry group might contain transformations that are not approximate symmetry transformations at all.

Again without going into detail, for a state space equipped with metric and an approximate symmetry group of it an *exact symmetry limit* is a metric for which that approximate symmetry group is the exact symmetry group. Since it is a physical system that is being considered, so that the metric has physical significance, an exact symmetry limit of it might correspond to a physically realizable system, to a conceivable but physically unrealizable system, or to nothing conceivable as a physical system at all. Returning to the nuclear physics example of maximalistic use of the symmetry principle presented in Section 5.5, we have charge symmetry as an approximate symmetry. An exact symmetry limit would have charge symmetry as an exact symmetry, i.e., the strong nuclear interaction would be acting, while the weak nuclear interaction, the electromagnetic interaction, and the Pauli exclusion principle would be "switched off." Such a situation is conceivable, since by describing it we conceive of it, but it certainly is not physically realizable. In terms of metric, any realistic metric for the actual situation must give nonzero "distances" for pairs of states differing only in that one or more protons in one are replaced by neutrons in the other and vice versa, since the weak nuclear interaction, the electromagnetic interaction, and the Pauli principle distinguish between such states. An exact symmetry limit for charge symmetry would give null "distance" for such pairs, expressing their equivalence in the absence of the weak nuclear interaction, the electromagnetic interaction, and the Pauli principle.

Another term for approximate symmetry is *broken symmetry*. The *symmetry-breaking factor* is whatever factor the (possibly hypothetical) switching off of brings about an exact symmetry limit. In our example the weak nuclear interaction, the electromagnetic interaction, and the Pauli principle constitute the symmetry-breaking factor. Another example is a crystal, which possesses broken displacement symmetry. The exact symmetry limit is an infinite crystal, obviously unobtainable in practice. It is the finiteness of the real crystal

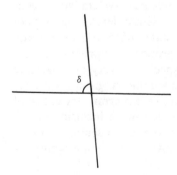

Fig. 6.1. System possessing approximate fourfold rotation symmetry for $\delta \neq 90°$ but $\delta \approx 90°$.

that breaks the displacement symmetry. Or consider the four-equal-straight-armed figure of Fig. 6.1, where each pair of opposite arms forms a straight line segment and $\delta \neq 90°$ but $\delta \approx 90°$, δ being the angle between the line segments, as shown. That system possesses approximate fourfold rotation symmetry. The symmetry-breaking factor is the difference between δ and $90°$. The exact symmetry limit is the system with $\delta = 90°$ and is realizable.

6.2. Spontaneous Symmetry Breaking

In discussions of symmetry and the symmetry principle, that the effect is at least as symmetric as the cause, the question of "spontaneous symmetry breaking" inevitably arises. There appear to be cases of physical systems where the effect simply has less symmetry than the cause, where the symmetry of the cause is possessed by the effect only as a badly broken symmetry, so that the exact symmetry group of the effect is a subgroup of the symmetry group of the cause, rather than vice versa. In other words, we seem to have broken symmetry with no symmetry-breaking factor. Among those cases are Garrett Birkhoff's (American mathematician, 1911–   ) hydrodynamic "symmetry paradoxes" [A4], and additional examples are noted later in this section. Yet, as we have convinced ourselves, the symmetry principle cannot but hold. The trouble is that what is taken to be the cause is not complete, and the complete cause is less symmetric. What is assumed to be the exact symmetry of the cause is really only an approximate symmetry, although to a very good approximation. The goodness of approximation is so good that we are deceived into believing that it is perfect and overlook the small symmetry-breaking factor that makes the symmetry only approximate. What is in fact usually overlooked is the influence of small, random fluctuations in physical systems. Thus the symmetry principle is saved. The exact symmetry of the cause remains, as it must, the minimal exact symmetry of the effect. But a good approximate symmetry of the cause can be possessed by the effect as a badly broken symmetry.

Just how do small, symmetry-breaking perturbations of a cause affect the symmetry of the effect? What can be said about the symmetry of an effect relative to the *approximate* symmetry of its cause? That depends on the actual nature of the physical system, on whatever it is that links cause and effect in each case. But we can consider the possibilities.

1. *Stability.* The deviation from the exact symmetry limit of the cause, introduced by the perturbation, is "damped out," so that the approximate symmetry group of the cause is the minimal symmetry group of the effect.

2. *Lability.* The approximate symmetry group of the cause is the minimal approximate symmetry group of the effect, of more or less the same goodness of approximation.

3. *Instability.* The deviation from the exact symmetry limit of the cause,

introduced by the perturbation, is "amplified," and the minimal symmetry of the effect is only the exact symmetry of the cause (including perturbation), with the approximate symmetry of the cause appearing in the effect as a badly broken symmetry. That is what is commonly called *spontaneous symmetry breaking*. To quote Birkhoff [A4].

> *Although symmetric causes must produce symmetric
> effects, nearly symmetric causes need not produce
> nearly symmetric effects*: a symmetry problem need
> have no *stable* symmetric solutions.

An example of a stable situation is the solar system. All the planetary orbits lie pretty much in the same plane, and that seems to have been the case for as long as observations have been recorded, or even for as long as the planets have existed, according to the modern theory of the origin of the solar system. That is a symmetry, reflection symmetry with respect to the plane of the solar system. Now if the state of the solar system at any instant is taken as the cause, its state at any later instant is an effect. Thus the solar system exhibits stability for reflection symmetry. In spite of the various and numerous internal (and external) perturbations that the solar system has suffered throughout its history, which might be expected to have broken the reflection symmetry more and more as time went on, the symmetry has obviously been preserved. (Symmetry in processes is discussed in detail in Chapter 7.)

As an example of lability, consider the DC circuit presented in Section 5.4 as an example of minimalistic application of the symmetry principle (Fig. 5.12). That system possesses twofold rotation symmetry (actually, as explained there, functional permutation symmetry). But it is the ideal system that is symmetric. No real circuit built according to that diagram will be exactly symmetric because of practical limitations on the accuracy of resistors, emf sources, and so on. Yet the effect, the currents and potential drops, always shows twofold rotation symmetry to reasonable accuracy. So the situation is certainly not one of instability. Is it stability or lability? To check that, we replace a fixed resistor by a variable one or a fixed emf source by a variable one and gradually break the symmetry of the cause, following the symmetry of the effect with ammeters and voltmeters as we do so. And we find that the symmetry of the effect gets broken as gradually as we break the symmetry of the cause. That proves that the system is labile for twofold rotation symmetry.

For an example of instability we return to the solar system, this time to its origin and evolution. Modern theory has the solar system originating as a rotating cloud of approximate axial symmetry and reflection symmetry with respect to a plane perpendicular to its axis. If that state of what is now the solar system is taken as the cause, the present state can be taken as the effect. And any axial symmetry the proto solar system once had has clearly practically disappeared during the course of evolution, leaving the solar system as we now observe it. The random, symmetry-breaking fluctuations in the original gas cloud grew in importance as the system evolved, until the original axial symmetry became hopelessly broken. Additional examples of instability are Birkhoff's "symmetry paradoxes," referred to previously.

An example exhibiting both stability and instability, although under different conditions, is a volume of liquid at rest in a container. Such a liquid is isotropic; its physical properties are independent of direction. That is a symmetry. The system is stable for isotropy, and our attempts to break the symmetry will be overcome by the liquid, if only it can. A tap on the side of the container introduces anisotropy, which is very soon damped out by the system, and isotropy is regained. A small crystal of the frozen liquid thrown into the liquid also breaks the symmetry, but it soon melts and isotropy returns. However, when the liquid is cooled to below its freezing point, the situation alters drastically. Let us imagine that it is cooled very slowly and evenly and that no isotropy-breaking perturbations are allowed; in other words, assume that the liquid is supercooled. It is then still in the liquid state and isotropic. If we now tap the container or throw in a crystal or otherwise introduce anisotropy, the supercooled liquid will immediately crystallize and thus become highly anisotropic. So in the subfreezing temperature range the system is unstable for isotropy; any anisotropic perturbation is immediately amplified until the whole volume becomes anisotropic and stays that way. If, in spite of our precautions, the liquid should spontaneously crystallize during or after cooling, the reason is that its own internal random fluctuations are sufficient to set off the instability. The cooler the liquid (below its freezing point), the greater its instability. The freezing point is the boundary between the temperature range of stability and that of instability. In fact, it can be defined that way.

Even in situations of lability and instability, if the perturbation is truly a random one, there is a sense in which the symmetry of the unperturbed cause, the exact symmetry limit, nevertheless persists as the minimal symmetry of the effect. That pivots on the nature of random perturbations, or, more accurately, on their definition. A random perturbation of an unperturbed cause is defined, for our purpose, as a fluctuating perturbation such that the symmetry of the *totality* of observed states of the perturbed cause tends to that of the unperturbed cause. In other words, a random perturbation should average out over all states of the perturbed cause to no perturbation at all. One might also express that in terms of observations of the state of the cause, where as the number of observations increases, the symmetry of the *totality* of observed states of the perturbed cause tends to that of the unperturbed cause. Thus, taking the totality of states of a cause as "the cause" in the symmetry principle, we have that for a random perturbation the totality of states of the effect, as "the effect" of the symmetry principle, tends to exhibit at least the symmetry of the unperturbed cause. Or, expressed in terms of observations, for a random perturbation, as the number of observations increases, the totality of observed states of the effect will tend to exhibit at least the symmetry of the unperturbed cause.

Fig. 6.2. Air current blowing against edge of wedge, viewed in cross section.

As an illustration consider the system consisting of a current of air blowing against the edge of a wedge, with the direction of the air current far before the wedge parallel to the bisector of the wedge, as in Fig. 6.2. Take as the cause the wedge and the incident air current a large distance before the wedge, and the air flow around the wedge as the effect. The cause possesses reflection symmetry with respect to the plane bisecting the angle of the wedge. So we expect the air flow above the wedge to be the mirror image of the air flow under the wedge. And, in fact, that is what happens, as long as the velocity of the incident air current is sufficiently low so that the flow is laminar. However, at sufficiently high velocities vortices form at the edge and are shed downstream. They are not produced symmetrically, but rather alternately at one side and the other. (That periodic phenomenon is the mechanism of production of "edge tones," the acoustic source of woodwind instruments.) The effect then apparently has lower symmetry than the cause, since random fluctuations in the direction of incident air flow make the reflection symmetry of the cause only an approximate symmetry. At low incident-flow velocities the situation is labile, and the resulting laminar flow around the wedge is approximately symmetric. (Underblowing a woodwind instrument produces no tone.) At high velocities the situation is unstable, and the vortical flow around the wedge lacks reflection symmetry altogether. But if the vortical flow is photographed sufficiently often at random times, the total collection of photographs will exhibit reflection symmetry; for every flow configuration photographed a mirror-image configuration will also appear among the pictures.

I would like to add that it seems to me that the recently developed "theory of catastrophes" [M67, M47, M61] should be a suitable mathematical language for the description of instabilities of the kind we have been discussing. See also [M28]. But since the mathematical theory was publicized with much fanfare in the 1970s, I have not become aware of any such extensive application of it.

The situation, discussed in Section 5.6, where the vacuum state of a quantum theory is less symmetric than the rest of the theory, can be viewed as broken symmetry, with the symmetry of the theory without the vacuum state taken as the exact symmetry limit and the vacuum state as the symmetry-breaking factor. However, such a situation is sometimes referred to as *spontaneous* symmetry breaking. That nomenclature is misleading, and the attribution of spontaneity is not justified. The vacuum state is an obvious component of the cause for such theories, and its effect in reducing the symmetry of the total cause from that of the exact symmetry limit is in no sense small, nor is it (or should it be) liable to be overlooked. What might justifiably be called *spontaneous* symmetry breaking in that connection is a higher-level theory, a super theory, that is supposed to explain the first theory, including the lower symmetry of the vacuum state compared with that of the rest of the theory, or at least it is supposed to explain the vacuum state. If the super explanation involves instability and amplification of perturbations, so that the vacuum state, as an effect of the super theory, comes out possessing a lower degree of symmetry than that of the unperturbed super cause, whatever that might be, then the use of the term "spontaneous" is reasonable and consistent with our previous discussion. A situation like that is described in Section 7.5.

6.3. Summary of Chapter Six

In Section 6.1 we introduced the idea of a metric in the state space of a physical system as a softening of the all-or-nothing character of an equivalence relation. That then allowed the definition of an approximate symmetry transformation as any transformation the maps every state to an image state that is "nearly" equivalent to the object state. We mentioned or briefly discussed the concepts of approximate symmetry group, goodness of approximation, an exact symmetry limit of an approximate symmetry group as a situation in which the approximate symmetry group is the exact symmetry group, broken symmetry as another term for approximate symmetry, and a symmetry-breaking factor as whatever factor the (possibly hypothetical) switching off of brings about an exact symmetry limit.

We discussed spontaneous symmetry breaking in Section 6.2. The crux of the matter was seen to be a question of stability: How do small, symmetry-breaking perturbations of a cause affect the symmetry of the effect? Although, by the symmetry principle, *exact* symmetry of a cause must appear in its effect, *approximate* symmetry of a cause might appear in the effect as exact symmetry, as approximate symmetry, or as badly broken symmetry. We saw examples.

CHAPTER SEVEN

Symmetry in Processes, Conservation, and Cosmic Considerations

In this chapter we discuss and formalize the symmetry of evolution and the symmetry of initial and final states in the analysis of the natural evolution of quasi-isolated systems into initial state and evolution. Symmetry of evolution is also known as symmetry of the laws of nature, and the discussion includes time-reversal symmetry. The considerations of symmetry of initial and final states lead to the equivalence principle and the symmetry principle for natural processes in quasi-isolated systems and to the general and special symmetry evolution principles for such systems. The latter two principles are both concerned with the nondecrease of the degree of symmetry during the evolution of quasi-isolated systems. We obtain an explanation for the empirical observation that macrostates of stable equilibrium of physical systems are often especially symmetric. Symmetry and entropy are shown to be related to each other. Then we discuss conservation and its relation to symmetry of the laws of nature. And finally we indulge in cosmic considerations about it all.

Chapters 4–6 are prerequisite. If you are on the "concept" track, a review of Sections 9.5, 10.2, and 10.3 is very useful for Sections 7.1, 7.2, and 7.3. The prerequisite scientific background for Chapters 5 and 6, with the addition of some kinetic theory and some familiarity with the concepts of macrostate and microstate, should suffice for the present chapter. Acquaintance with the "big bang" and some elementary particle physics should also prove useful. The prerequisite mathematics for Section 7.4 is calculus, including partial differentiation.

7.1. Symmetry of the Laws of Nature

First a paragraph for the "concept" trackers: "Symmetry of the laws of nature" is another term for symmetry of evolution of quasi-isolated systems, which is discussed in Section 9.5. *Laws of nature*, or *laws of evolution*, refer to the lawful behavior of quasi-isolated systems that is found when such systems are analyzed into initial state and evolution, as discussed in Section 10.2. Laws are descriptions of nature's order and are thus expressions of what in

Section 5.1 we call "causal relations in systems." But in the present case the term "systems," which in Chapters 4 and 5 is intentionally very vague and general, refers specifically to natural processes of quasi-isolated systems. The whole evolution process is the "system." Initial and final states are "parts" of that system. In fact, for lawful behavior of quasi-isolated systems the initial state is the "cause subsystem" and the final state the "effect subsystem," as is explained in Sections 9.5 and 10.2.

Now for all readers: *Laws of nature*, or *laws of evolution*, are what in Chapter 5 we call "causal relations in systems" when the "systems" are processes in physical systems, with the "cause subsystem" being the initial state of the physical system and the "effect subsystem" its final state. Please note well and beware of the two senses in which the term "system" is being used here! One sense is that of a physical system, which can be described in terms of initial and final states and natural evolution from the former to the latter. The other is "whatever we investigate the properties of," according to the presentation in Section 4.1. And here we are investigating the properties of the evolution processes of quasi-isolated systems (in the former sense), where those processes involve initial states evolving into final states. So in the latter sense of the term "system" the initial state, the final state, and the evolution are all parts of the system that is the process.

Yes, it is the dynamic process, the temporal evolution of the physical system, that is taken as the "system", while the state of the physical system at any initial time is the "cause subsystem" and the state at any final time is the "effect subsystem." The laws of nature, or laws of evolution, are the causal relation between cause and effect subsystems in such cases, the causal relation between initial and final states of physical processes. Equivalently, and in more familiar language, the laws of nature can be viewed as the natural temporal development of physical systems from initial states to final states. See [M19] and [M52].

It must be emphasized that we are considering only quasi-isolated physical systems, where by "quasi-isolated" we mean that there is minimal interaction with the rest of the world, that the physical systems evolve, to the extent possible, under internal influences only. That is implied by the existence of causal relation, cause subsystem, and effect subsystem, with initial states leading to unique final states.

We have all heard of various symmetries and approximate symmetries that are ascribed to the laws of nature, such as special relativistic symmetry, $SU(3)$ symmetry, isospin symmetry, charge symmetry, particle–antiparticle-conjugation (C) symmetry, time-reversal (T) symmetry, space-inversion (P) symmetry, CP symmetry, CPT symmetry. (We cannot go into a description of them here.) There are two points of view about the meaning of symmetry of the laws of nature.

1. A scientist (with a laboratory) investigates nature and discovers laws. Another scientist (with his or her laboratory), related to the first by some

transformation, also investigates nature and discovers laws. If these laws are the same as those discovered by the first scientist (or if a certain subset of these is the same as a certain subset of those), and if, moreover, that is true for all pairs of scientists related by the transformation, then the transformation is a symmetry transformation of the laws of nature (or of a certain subset of them). This is called the passive point of view.

2. If among all processes that can be conceived as occurring naturally (or a certain subset of them) any pair related by a certain transformation are either both allowed or both forbidden by nature (or, in quantum phenomena, both have the same probability), then that transformation is a symmetry transformation of the laws of nature (or of a certain subset of them). This is called the active point of view.

For example, the symmetry of the special theory of relativity can be expressed passively: (1) All pairs of scientists with arbitrary spatiotemporal separation, arbitrary relative orientation, and arbitrary constant rectilinear relative velocity discover the same laws of nature, including the speed of light. Or it can be put actively: (2) Among all processes that can be conceived as occurring naturally, including the propagation of a light signal over a certain distance during a certain time lapse, any two that are identical except for arbitrary spatiotemporal separation, arbitrary relative orientation, and arbitrary constant rectilinear relative velocity are either both allowed or both forbidden by nature (or have the same probability).

Another example, *CP* symmetry, which seems to be valid in all of nature except for a certain class of weak or superweak interactions among elementary particles, can be formulated passively: (1) All pairs of scientists that are particle–antiparticle conjugates and space-inversion images of each other discover the same laws of nature (with certain exceptions). The transformation involved here is converting a scientist into an antiscientist, i.e., replacing all the protons, neutrons, electrons, and other particles comprising the scientist and his or her laboratory with antiprotons, antineutrons, positrons, and other antiparticles, respectively, and then point inverting the antiscientist (or first invert and then conjugate; the results are the same). Since antiscientists are not at present physically realizable (at least not in the sense just defined), the passive point of view seems absurd, so one might prefer the active one: (2) Among all processes that can be conceived as occurring naturally (with certain exceptions), any two that are identical except for particle-antiparticle conjugation and space inversion are either both allowed or both forbidden by nature (or have the same probability).

The two points of view are called "passive" and "active," respectively, because of the nature of the transformations under which the laws of nature are symmetric. A transformation is called "passive" if it does not change the world but changes the way the world is observed, i.e., it changes the frame of reference. For a geometric transformation, for example, that would be a change of coordinate system. A transformation is called "active" if it does not change the observer but changes the rest of the world. The same frame of reference is used for a changed world. In the geometric case the locations of events would be changed with respect to a fixed coordinate system.

I prefer the active (second) point of view for considering symmetry of the laws of nature. The main reason is that, as we saw in the example concerning *CP* symmetry, transformed frames of reference (observers) are not always physically realizable. From the active point of view a symmetry transformation of the laws of nature can be concisely defined, just as in Section 10.2, as a transformation for which the transformed result of an experiment is the same as the result of the transformed experiment for all experiments (or for some subset of experiments, in which case we have a symmetry transformation of a subset of the laws of nature).

Here is a precise, diagrammatic formulation of symmetry of the laws of nature. (It is the expression, in the symmetry formalism we are developing, of the qualitative formulation presented in Section 10.2.) The laws of nature, or laws of evolution, are expressed by a transformation, denoted by N (for "nature"), of the state space of every quasi-isolated physical system into itself, a mapping giving as the image of any state the final state evolving naturally from the object state as initial state and thus exhibiting the causal relation between initial and final states. See Fig. 7.1. (Since not all states are necessarily obtainable as final states, there may be states that do not serve as images in that mapping. How then, you may well ask, are such states obtained to serve as initial states? They are set up by tampering with the system; i.e., they are the final states of processes during which the system is *not quasi-isolated*!) So if u denotes the initial state of any physical system, $N(u)$ is the final state that is the result of the evolution of the system from initial state u. We can also use the notation

$$u \xrightarrow{N} N(u)$$

to make the transformation graphically look more like a process. That dis-

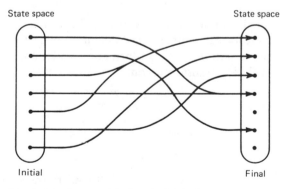

State space   State space

Initial   Final

Fig. 7.1. Laws of evolution for quasi-isolated physical system are mapping of system's state space into itself, where image of any state is final state evolving from object state as inital state.

crete, initial → final model of evolution will best serve the purposes of our discussion. Continuous evolution is obtained in the limit of infinite reiteration of infinitesimal discrete steps.

Let $\mathcal{T}$ denote any transformation of every physical system, such as rotation, spatial displacement, reflection, particle–antiparticle conjugation. The image of state u of any physical system is state $\mathcal{T}(u)$ of the same system, or

$$u \xrightarrow{\mathcal{T}} \mathcal{T}(u),$$

although we are not indicating a process this time.

The process

$$u \xrightarrow{N} N(u)$$

is (or could be) the running of an experiment, where u denotes the initial experimental setup and $N(u)$ denotes the result. Transforming the experimental result $N(u)$ by $\mathcal{T}$ gives image state $\mathcal{T}N(u)$, or

$$N(u) \xrightarrow{\mathcal{T}} \mathcal{T}N(u).$$

Now consider the transformed experiment, the image of the above experiment under $\mathcal{T}$. Its initial setup is obtained by transforming state u to state $\mathcal{T}(u)$, or

$$u \xrightarrow{\mathcal{T}} \mathcal{T}(u).$$

Starting from initial state $\mathcal{T}(u)$, the experiment takes place, proceeding according to the laws of nature, and yields the result $N\mathcal{T}(u)$, or

$$\mathcal{T}(u) \xrightarrow{N} N\mathcal{T}(u).$$

Combining those results, our definition of a symmetry transformation of the laws of nature, a transformation $\mathcal{T}$ for which for all experiments the transformed result is the same as the result of the transformed experiment (see Section 10.2), now becomes

$$\mathcal{T}N(u) = N\mathcal{T}(u)$$

for all states u, or diagrammatically as in Fig. 7.2. Since that holds for all states u, we have the formal, mathematical definition of a symmetry transformation of the laws of nature: The transformation under consideration commutes with the evolution transformation,

$$\mathcal{T}N = N\mathcal{T}$$

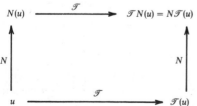

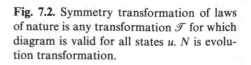

Fig. 7.2. Symmetry transformation of laws of nature is any transformation $\mathcal{T}$ for which diagram is valid for all states u. N is evolution transformation.

(see Section 4.2). (For a symmetry transformation of a subset of the laws of nature things must be formulated accordingly in terms of all states u belonging to a certain subspace of state space.)

The fundamental point underlying those definitions and formulations (as is discussed in Section 10.2) is this: A symmetry of the laws of nature is an indifference of the laws of nature. For a transformation to be a symmetry transformation of the laws of nature, the laws of nature must ignore some aspect of physical states, and the transformation must affect that aspect only. A pair of initial states related by such a transformation are treated impartially by the laws of nature; they evolve into a pair of final states that are related by precisely the same transformation. The laws of nature are blind to the difference between the two states, which is then preserved during evolution and reemerges as the difference between the two final states.

As an example, consider spatial-displacement symmetry of the laws of nature, meaning that the laws of nature are the same everywhere. If we perform two experiments that are the same except for one being here and the other being there, they will yield outcomes that are the same except for one being here and the other there, respectively. And that is found to be valid for all experiments and for all heres and theres. For a picture see Fig. 7.3.

Those insensitivities of the natural evolution of quasi-isolated systems remind us of the "impotencies" of scientific laws, discussed in Section 5.2. Indeed, since laws are expressions of nature's order, all symmetries of evolution must appear as impotencies of laws, if the laws are to express nature's order faithfully. Any subset of natural evolution processes will possess the symmetry of the full set and might have additional symmetry. Partial laws, those laws that are concerned with only part of the phenomena of nature, might accordingly possess additional impotencies.

For example, since natural evolution is symmetric under spatial displacements, no scientific law may accept absolute position as input. For examples of partial theories, the gas laws are concerned with macroscopic states of gases, and Kirchhoff's laws are concerned with electric currents. In the former case, although the laws of nature do have something to say about which

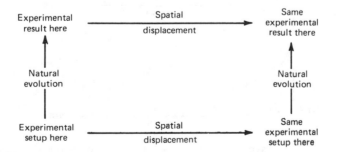

Fig. 7.3. Spatial-displacement symmetry of laws of nature. Diagram is valid for all experiments and for all heres and theres.

gas molecules go where and when, the gas laws ignore microscopic aspects of the gas. In the latter case, although the laws of nature do apply to the motion of each of the moving electrons comprising the electric current, Kirchhoff's laws ignore such details.

The set of invertible symmetry transformations of the laws of nature, i.e., the set of all invertible transformations commuting with the evolution transformation, is easily shown to form a group. We call that group *the symmetry group of the laws of nature*.

Another possible symmetry of the laws of nature, very different in kind from the others, is temporal-inversion, or time-reversal, symmetry, also called reversibility. The time-reversal transformation acts on a process by replacing each state by its time-reversal image and reversing the temporal order of events. A moving picture projected in reverse is a good model of time reversal. The time-reversal image of a state depends on the type of state involved; for classical mechanics it is obtained merely by reversing the senses of all velocity vectors. The question then is whether the image process, leading from the time-reversed final state to the time-reversed initial state, is the process that would evolve naturally from the time-reversed final state. If that is so for all processes, the laws of evolution possess time-reversal symmetry, or are reversible. That is illustrated in Fig. 7.4.

In symbols the object process is

$$u \xrightarrow{N} N(u).$$

The time-reversal image of the final state $N(u)$ is $TN(u)$,

$$N(u) \xrightarrow{T} TN(u),$$

where T stands for time reversal. The natural evolution from state $TN(u)$ is

$$TN(u) \xrightarrow{N} NTN(u).$$

And if the final state of this process is the time-reversal image of the initial

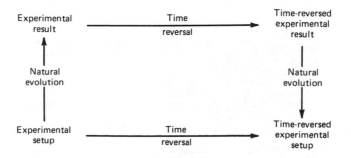

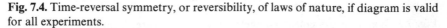

Fig. 7.4. Time-reversal symmetry, or reversibility, of laws of nature, if diagram is valid for all experiments.

state of the object process, i.e., if

$$NTN(u) = T(u)$$

for all states u, we have time-reversal symmetry, the definition of which is then

$$NTN = T.$$

Note how that differs from the definition, derived previously, of a symmetry transformation of the usual kind.

With a little thought it will be clear that for the laws of nature to possess time-reversal symmetry, i.e., for the definition we just derived to be valid, it is necessary (although not sufficient) that the evolution transformation be both one-to-one and onto, that N be invertible. So, if we like, we can put the definition we just obtained in the form

$$TN = N^{-1}T.$$

At present time-reversal symmetry, or reversibility, seems to be valid in nature for the laws of evolution of almost all microscopic systems, more precisely, for all microscopic systems that do not contain neutral kaons. On the other hand, it is invalid for almost all macroscopic systems.

Note that time reversal has nothing to do with processes "running backwards in time," and nothing we have said about time reversal or time-reversal symmetry of the laws of nature should be construed as implying anything like a "backward flow of time." All evolutions we are considering, even the evolution from the time-reversed final state, are natural evolutions. The use of the inverse N^{-1} of the natural evolution mapping N in the form $TN = N^{-1}T$ of the symbolic definition of time-reversal symmetry of the law of evolution is but a formal device and does not indicate "time flowing backwards." True, we often represent time by a coordinate axis labeled t, and that is indeed very useful. However, it is then all too tempting to spatialize time and imagine that everything is "moving" along that axis in the positive t direction. So then why not "move" in the negative t direction? Well, that is just not the nature of time. But we are going astray.

PROBLEMS

1. No real physical system is perfectly isolated. Discuss possible kinds of external influence. Consider the question of stability against external influence. How is it that we discover laws of nature, although our experimental systems are not precisely isolated? Speculate about what might happen in perfectly isolated physical systems.

2. Prove that the set of invertible symmetry transformations of the laws of nature forms a group.

3. What underlies the difference between time-reversal symmetry and symmetries whose symmetry transformations commute with the evolution transformation?

7.2. Symmetry of Initial and Final States, the General Symmetry Evolution Principle

In the analysis of the behavior of quasi-isolated systems into initial state and evolution, the laws of evolution, or the laws of nature, do not exhaust all of what is going on. The laws of nature do indeed determine how any process, once started, will evolve and what its outcome will be. But it is the initial state, the situation at any given single instant, that determines just which process is to take place. And, with an important exception discussed in Section 7.5, the laws of nature do not determine initial states. Or perhaps we might say that initial states are whatever nature allows us to have control over, at least in principle, or, perhaps even better, whatever nature prefers not to be bothered with.

For an example in classical mechanics, consider a set of bodies interacting only with each other. The laws of evolution are Newton's laws of motion (Isaac Newton, English philosopher and mathematician, 1642–1727) and the forces among the bodies. We may arbitrarily specify the positions, momenta, orientations, and angular momenta of all the bodies at a given time but no more than that. So those specify initial states.

Since the laws of nature are beyond our control, the initial state uniquely determines for us the process and its outcome—the final state of the quasi-isolated system. Thus, as described in the preceding section, initial state and outcome are in causal relation. Taking the whole process as our "system," the initial state and the final state are cause and effect subsystems, respectively.

I again emphasize that we are considering only quasi-isolated physical systems, in which initial states do indeed uniquely determine the outcomes of processes, in which there is a minimum of outside influences messing things up.

(For "concept" trackers: We now refer to the discussion of Section 10.3 and express in the symmetry formalism we are developing the idea of symmetry of states.) Symmetry of the laws of nature (time-reversal symmetry excluded) determines an equivalence relation in the state space of a system: A pair of states is equivalent if and only if they are indistinguishable by the laws of nature, i.e., if and only if some element of the symmetry group of the laws of nature carries one into the other. Thus the symmetry group of the laws of nature determines a decomposition of state space into equivalence subspaces, the physical significance of which is that the members of each equivalence subspace are all those states and only those that are indistinguishable by the laws of nature. Then, following reasoning similar to that of Section 5.2, we obtain *the equivalence principle for processes* in quasi-isolated systems:

> *Equivalent states, as initial states,*
> must *evolve into equivalent states, as*
> *final states, while inequivalent states may*
> *evolve into equivalent states.*

See Fig. 7.5.

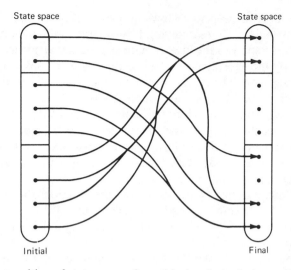

Fig. 7.5. Decomposition of state space of quasi-isolated physical system into equivalence subspaces determined by symmetry group of laws of nature. According to the equivalence principle equivalent states always evolve into equivalent states, while inequivalent states may also do so.

That principle is valid and useful for natural evolution processes, for any subset of them, for universal scientific laws, or for partial laws.

For an example of the equivalence principle for processes, since the laws of nature possess spatial-displacement symmetry, initial states that differ only in position evolve into final states that differ only in position.

Or all known laws of nature are symmetric under interchange of elementary particles of the same species, so such particles are, as far as we know, inherently indistinguishable. Two initial states differing only in the interchange of, say, two electrons cannot, by any known process, lead to distinguishable final states.

Or certain laws of nature might apply to macroscopic states of a system and be indifferent to microstates, as long as the latter correspond to the same macrostate. Then two different initial microstates corresponding to the same initial macrostate will evolve into final microstates that correspond to the same final macrostate. Among those are the gas laws. Kirchhoff's laws care nothing about the composition and structure of resistors and emf sources, as long as they have the resistance and emf they are supposed to have. Newton's laws are obviously indifferent to many aspects of the states to which they apply (color, odor, etc.).

Let us call the equivalence relation defined by the symmetry group of the laws of nature "initial equivalence" for convenience. That really is for convenience in the following discussion, and is not just to use two words where one will do. We now define another, possibly different, equivalence relation, which we call "final equivalence," as follows. A pair of states is final-equivalent if and only if the pair of states evolving from them is initial-equivalent, i.e., is equivalent with respect to the laws of nature. Initial equivalence implies final equivalence, by the equivalence principle for processes, since initial-equivalent state must evolve into initial-equivalent states and are thus final-equivalent. However, states that are not initial-equivalent may also evolve into initial-equivalent states and thus be final-equivalent. So, according to the equivalence principle for processes, in the decomposition of the state space of a physical system into final-equivalence subspaces, each final-equivalence subspace consists of one or more initial-equivalence subspaces in their entirety. See Fig. 7.6.

When considering symmetry in processes, the process is the "system" in the general sense of the symmetry and equivalence principles, and the set of all possible processes of a physical system is the "state space" in the same general sense. Transformations of that "state space" act on processes. However, as we have seen, the set of all possible processes stands in one-to-one correspondence with the set of states of the physical system, since every state, as initial state, initiates a unique process, and every process starts with some initial state. So, to keep things as uncomplicated as possible (relatively speaking, of course), we apply our transformations to the physical system's state space rather than to process space. As we transform the states about, we keep in mind that each one drags around with it the process it initiates, just like mice running around in a cage, each carrying its tail.

Then since, as we saw above, initial states are the cause subsystem and final states the effect subsystem for processes in quasi-isolated physical systems, we have that the symmetry group of the cause (see Section 5.3) is the group of all invertible transformations of state space that preserve initial-equiva-

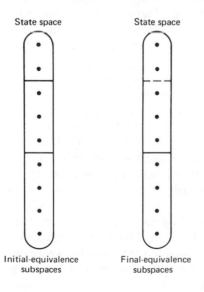

State space   State space

Initial-equivalence   Final-equivalence
subspaces   subspaces

Fig. 7.6. Decomposition of state space of quasi-isolated physical system into initial-equivalence subspaces and into final-equivalence subspaces for example of Fig. 7.5. Each final-equivalence subspace consists of one or more initial-equivalence subspaces.

lence subspaces. And the symmetry group of the effect is the group of all invertible transformations preserving final-equivalence subspaces. And since, as follows from the discussion two paragraphs above, any transformation that preserves initial equivalence must also preserve final equivalence, we have the result:

> The "initial" symmetry group (that of the cause)
> is a subgroup of the "final" symmetry group
> (that of the effect).

That is *the symmetry principle for processes* in quasi-isolated physical systems. In that sense we can say:

> For a quasi-isolated physical system the degree of
> symmetry cannot decrease as the system evolves,
> but either remains constant or increases.

(For degree of symmetry see Section 4.6.) We call that result *the general symmetry evolution principle*. The adjective "general" is included in the name of the principle because the principle is derived from very fundamental considerations with no additional assumptions. Thus it is indeed general, so general, in fact, as to make it rather useless, as we will explain and remedy in the next section. Our motive for deriving it is theoretical, not utilitarian; we should see where our fundamental considerations lead us when applied to the evolution of isolated physical systems.

PROBLEM

We stated that states differing only in the interchange of two elementary particles of the same species are indistinguishable by the laws of nature. And, indeed, we can think of no way of distinguishing among such states. However, due to the spatial-displacement symmetry of the laws of nature, states differing only in their location are also indistinguishable by the laws of nature. Is there a way of distinguishing among such states? If so, consider the lack of analogy with the former case. Are we not running into difficulty? Try to resolve the difficulty.

7.3. The Special Symmetry Evolution Principle and Entropy

The reason the general symmetry evolution principle, although perfectly valid, is not very useful is that, when we consider the evolution of a quasi-isolated physical system, we consider the sequence of states it passes through and their symmetry and are usually not interested in the entire state space of the system, in terms of whose transformations the "initial" and "final" symmetry groups are defined in the preceding section. The symmetry group of a single state is the group of all invertible transformations of the physical system that carry the state into equivalent states, where equivalence is with respect to the laws of evolution. Thus the symmetry group of a state is just the group of permutations of all members of the equivalence subspace in state space to which the state belongs (where we generalize suitably for equivalence subspaces with discretely or continuously infinite populations, or, to use the terms presented in Section 2.1, of denumerably or nondenumerably infinite orders). Let us recall that by "equivalence subspace in state space" we mean the equivalence class of states that are indistinguishable by the laws of nature, as explained in the previous section. So, referring to the discussion in Section 4.6, the degree of symmetry of a state can be measured by the population of its equivalence subspace: The more states equivalent to it, the higher its degree of symmetry.

What can be stated about the degrees of symmetry of the sequence of states through which a quasi-isolated system evolves? Or equivalently, what do we know about the populations of the sequence of equivalence subspaces to which those states belong? In spite of the general symmetry evolution principle we know nothing about that in general. We can, however, obtain a result by making the assumption of nonconvergent evolution: Different states always evolve into different states. Then the population of the equivalence subspace of a final state is at least equal to that of the initial state that evolved into the final state, since, by the equivalence principle, all members of the initial state's equivalence subspace evolve into members of the final state's equivalence subspace and, by the nonconvergence assumption, the number of the latter equals the number of the former. Moreover, additional states, inequivalent to the initial state, may also evolve into members of the final state's equivalence subspace (convergence of equivalence subspaces), and these members will be distinct from those we just counted. Thus:

As a quasi-isolated system evolves, the populations of the equivalence subspaces (equivalence classes) of the sequence of states through which it passes cannot decreases, but either remain constant or increase.

Or, equivalently,

The degree of symmetry of the state of a quasi-isolated system cannot decrease during evolution, but either remains constant or increases.

See [M48]. We call that result *the special symmetry evolution principle*. See Fig. 7.7. (Compare the special and the general symmetry evolution principles; both derive from the equivalence principle for processes, while the special principle involves the additional assumption of nonconvergent evolution.)

The symmetry group of the final state, the group of permutations of all members of its equivalence subspace, clearly includes as a subgroup the symmetry group of the initial state, the permutation group of *its* equivalence subspace (all groups being considered abstractly). Actually it includes as subgroups the symmetry groups of all states evolving into states equivalent to

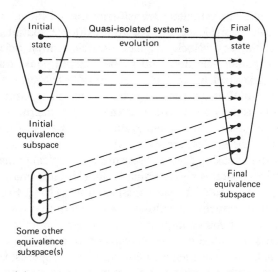

Fig. 7.7. The special symmetry evolution principle. Population, measuring degree of symmetry, of final equivalence subspace at least equals that of initial equivalence subspace. That follows from the equivalence principle and assumption of nonconvergent evolution. Solid arrow indicates actual process being considered, while dashed arrows indicate other possible processes.

the final state. Furthermore, it even includes as a subgroup the direct product of the permutation groups of all distinct equivalence subspaces of states evolving into states equivalent to the final state (or, we could say, of all distinct equivalence subspaces converging to the final equivalence subspace).

The assumption of nonconvergent evolution (of states, of course, not of equivalence subspaces) seems to be valid in nature for microscopic processes, at least as far as we now know. However, it is not valid for macroscopic processes, which is just fine, since the essence of the macroscopic character of a macrostate is that a macrostate is an equivalence subspace of microstates, and such equivalence subspaces are just what we are dealing with here.

Evolution with constant degree of symmetry is typical of microscopically considered systems, systems about which sufficient information is available so that they need not be treated by statistical methods. The mechanism of evolution with constant degree of symmetry is, of course, nonconvergence of equivalence subspaces. Reversibility, or time-reversal symmetry, of the laws of evolution implies evolution with constant degree of symmetry. (If the degree of symmetry increased, the time-reversed process would take place with decreasing degree of symmetry, which is forbidden by the special symmetry evolution principle.) Thus evolution with constant degree of symmetry is a necessary (but not sufficient) condition for reversibility.

As an example of evolution with constant degree of symmetry, consider a system consisting of a few nucleons and pions about which sufficient information is available. The symmetry transformations of the laws of evolution, which define state equivalence, are for this example spatial and temporal displacements, rotations, velocity boosts, particle–antiparticle conjugation, spatial (point or plane) reflection, and global isospin phase transformations (which are not described here). The evolution of that system is reversible, since its laws of evolution are time-reversal symmetric (quantum time-reversal symmetry). Thus the system evolves with constant degree of symmetry. As it evolves, the populations of the equivalence subspaces of the sequence of states constituting the process are all equal.

Evolution with increasing degree of symmetry is typical of macroscopically considered systems, systems about which insufficient information is available so that they must be treated by statistical methods [M4]. From the special symmetry evolution principle it follows immediately that evolution with increasing degree of symmetry implies irreversibility, or time-reversal asymmetry, of the laws of evolution. The mechanism of evolution with increasing degree of symmetry is convergence of equivalence subspaces, i.e., convergent evolution of macrostates. See Fig. 7.8. So, thinking macroscopically of macrostates, macrostate spaces, and macroscopic evolution transformations N, such N's are not invertible.

An example of evolution with increasing degree of symmetry is a confined, quasi-isolated gas considered macroscopically. Let the gas initially occupy any part of the container and have volume V_i and temperature θ_i. The final macrostate is a homogeneous distribution of gas in the whole container with volume V_f and temperature θ_f. All initial macrostates with the same V_i and θ_i (but with different parts of the container occupied) lead to the same final macrostate (and there are also initial macrostates with different V_i and θ_i that lead to the same final macrostate as well). The macroscopic laws of evolution are blind to the differences among microstates corresponding to the same macrostate. Thus each macrostate is an equivalence subspace of microstate space. In microscopic terms there is convergence of equivalence subspaces during evolution; not only do equivalent initial microstates (corresponding to the same V_i and θ_i and the same occupied part of the container) lead to equivalent final microstates, but there are also inequivalent initial microstates (corresponding to the same V_i and θ_i but different occupied parts of the

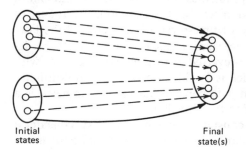

Initial
states

Final
state(s)

Fig. 7.8. Convergent evolution of macrostates (equivalence subspaces of microstates). Circles denote microstates, ellipses denote macrostates. Dashed arrows represent (nonconvergent) evolution of microstates, while solid arrows represent convergent evolution of macrostates.

container, or to different V_i and θ_i) that lead to equivalent final microstates. Since microstates do not converge during evolution, the population of the final equivalence subspace, which is the number of microstates corresponding to the final macrostate, is larger (actually very much larger) than the population of the initial equivalence subspace.

The final macrostate in this example is a state of stable equilibrium of the system. As such it is the outcome of evolution from a very large number of initial macrostates. By the special symmetry evolution principle its symmetry group (the group of permutations of all microstates corresponding to it) must include as a subgroup the direct product of the symmetry groups of all those initial macrostates, as we observed above, and must therefore be of relatively high order. In other words:

> *The degree of symmetry of a macrostate of*
> *stable equilibrium must be relatively high.*

That is a general theorem, independent of the example in the context of which it was derived. Although that symmetry is with respect to permutations of equivalent microstates, it can have macroscopic manifestations. In the example the final state is homogeneous, i.e., symmetric under all permutations of subvolumes of the gas. The theorem is in accord with observation [M64].

But what about those systems that insist on evolving toward reduced symmetry in spite of our theoretical arguments? For example, the solar system is thought to have evolved from a state of axial and reflection symmetry to the present, less symmetric state of only reflection symmetry. And the evolutionary development of plants and animals seems to be toward a reduction of symmetry. The answer is that those systems do not fulfill the conditions of our argument. They are either only approximately symmetric (while our argument is based on exact symmetry) or not isolated (i.e., at best quasi-isolated, thus possibly subject to external influences, which we exclude) or both and are not stable against such perturbations. Since no symmetry of a macroscopic physical system is exact and no physical system is absolutely isolated (both conditions being convenient and often useful idealizations), the crucial point is that of stability; a sufficiently stable system will behave as we described, will evolve with nondecreasing symmetry, and, if macroscopic, will have relatively highly symmetric states of stable equilibrium.

Under the correspondence

$$\text{degree of symmetry} \leftrightarrow \text{entropy}$$

the special symmetry evolution principle and the second law of thermodynamics are isomorphic. Both are concerned with the evolution of quasi-isolated systems; both state that a quantity, a function of macrostate, cannot decrease during evolution. So, following good science practice, we assume that a fundamental relation exists between entropy and degree of symmetry and look for a functional relation giving the value of one as a strictly monotonically ascending function of the other. That is already known, however, from the statistical definition of entropy,

$$S = k \log W,$$

where W is the number of microstates (the population of the equivalence subspace of microstate space) corresponding to the macrostate for which the value of the entropy, S, is thus defined. (k is the Boltzmann constant. That definition holds for the case of nonconvergence of microstates during evolution.) As mentioned, W can be used to measure the degree of symmetry of the macrostate, and the function $k \log W$ is indeed a strictly monotonically ascending function of W. See [M56].

Thus entropy and degree of symmetry for quasi-isolated systems either remain constant together or increase together, with a concomitant constancy or decrease in the degree of order for the system. Note that *disorder* stands in positive correlation with symmetry. Total disorder is maximal symmetry. Organization implies and is implied by reduction of degree of symmetry. The hypothetical "heat death" or "entropy death" of the Universe can also be called its "symmetry death."

PROBLEM

The subject of the last paragraph of this section deserves further contemplation, I think. Consider various implications of it. Compare with Pierre Curie's (French physicist and chemist, 1859–1906) article [A17], where he states that it is asymmetry ("dissymétrie," to use his terminology) that produces phenomena.

7.4. Conservation

There is more to be said about symmetry of the laws of nature, and that concerns conservation. As you most likely know, there are a number of conservations, also called "conservation laws," that hold for quasi-isolated systems. The best known of them are conservation of energy, conservation of linear momentum, conservation of angular momentum, and conservation of electric charge. What is meant is that, if the initial state of any quasi-isolated physical system is characterized by having definite values for one or more of those quantities, then any state that evolves naturally from that initial state will have the same values for those quantities. See [M18].

For example, in a particle scattering experiment the vector sum of the linear momenta of all the particles before the scattering has taken place equals the vector sum of the linear momenta of all the particles after they have interacted with each other. In addition, the total charge of the particles before collision equals their total charge afterward. And similarly for total angular momentum and total energy. Those conservations hold even if particles are produced or annihilated during the process!

It turns out that each conservation is intimately and fundamentally related to a group of symmetry transformations of the laws of nature. Although there is considerable theoretical understanding of that relation, there is still much room for further investigation of it. A theoretical discussion of that relation is beyond the scope of this book. However, let us mention the symmetries of the laws of nature that are related to the conservations referred to above and show (for three of the four) how each conservation can be derived analytically from its related symmetry for a simple mechanical system.

Conservation of energy is related to symmetry of the laws of nature under the group of temporal-displacement transformations, to the fact that the laws of nature do not change with time. That is called temporal homogeneity of the laws of nature. It means that for every naturally allowed process all processes that are identical with that one except for their occurring at different times are also allowed by nature.

Consider the nonrelativistic system of a single point particle of mass m moving in one dimension in a potential V that is a function only of the particle's coordinate x. Since the potential has no explicit dependence on time t, the laws governing this system's evolution are temporal-displacement symmetric. Denote the particle's velocity by $\dot{x}$ and its acceleration by $\ddot{x}$. Its total energy is

$$E = \tfrac{1}{2}m\dot{x}^2 + V.$$

The time rate of change of the total energy is

$$\frac{dE}{dt} = m\dot{x}\ddot{x} + \frac{dV}{dx}\dot{x}.$$

The derivative dV/dx equals, by the definition of potential, the negative of the force on the particle, which in turn, by Newton's second law of motion, equals $-m\ddot{x}$, giving

$$\frac{dE}{dt} = m\dot{x}\ddot{x} - m\ddot{x}\dot{x} = 0.$$

Thus the total energy does not change with time; it is conserved.

Conservation of linear momentum is related to symmetry of the laws of nature under the group of spatial-displacement transformations, to the fact that the laws of nature are the same everywhere. That is called spatial homogeneity of the laws of nature. It means that for every naturally allowed process all processes identical to that one but taking place at different locations are also allowed by nature.

Consider a nonrelativistic one-dimensional system consisting of a number of point particles, labeled by i, j, of masses m_i. Let x_i denote the coordinate of the ith particle, $\dot{x}_i$ its velocity, and $\ddot{x}_i$ its acceleration. Let the particles interact via a potential that depends only on the differences of the particles' coordinates, $x_i - x_j$. That potential is independent of the location of the system relative to the coordinate origin, so the laws governing the system's evolution are spatial-displacement symmetric. The total linear momentum of the system is

$$P = \sum_i m_i \dot{x}_i.$$

The time rate of change of the total momentum is

$$\frac{dP}{dt} = \sum_i m_i \ddot{x}_i.$$

The term $m_i \ddot{x}_i$ equals, according to Newton's second law, the force on the ith particle, which equals $-\partial V / \partial x_i$ by the definition of potential. Thus

$$\frac{dP}{dt} = -\sum_i \frac{\partial V}{\partial x_i}$$

$$= -\sum_i \sum_{j \neq i} \frac{\partial V}{\partial (x_i - x_j)}$$

$$= 0,$$

since each pair of terms with i and j interchanged cancel in the double sum. Thus the total momentum does not change with time; it is conserved.

Conservation of angular momentum is related to symmetry of the laws of nature under the group of all rotations about all axes through a point, to the fact that the laws of nature are the same in all directions. That is called isotropy of the laws of nature. It means that for every naturally allowed process all processes that are identical with that one but have different spatial orientations are also allowed by nature.

Consider the nonrelativistic system of a single point particle of mass m moving in a plane in a central potential, i.e., a potential depending only on the particle's distance from the origin. Since that potential is independent of the particle's orientation relative to the coordinate axes, the laws govening the system's evolution are rotation symmetric about the axis through the origin and perpendicular to the plane of the system. Let (x, y) denote the particle's coordinates. Then its distance from the origin is

$$r = \sqrt{x^2 + y^2}.$$

Denote the x-component of the particle's velocity by $\dot{x}$, the y-component of its velocity by $\dot{y}$, the x-component of its acceleration by $\ddot{x}$, and the y-component of its acceleration by $\ddot{y}$. The angular momentum of the particle with respect to the origin is

$$M = m(x\dot{y} - y\dot{x}).$$

The time rate of change of the angular momentum is

$$\frac{dM}{dt} = m(x\ddot{y} - y\ddot{x}).$$

The expression $m\ddot{y}$ equals, by Newton's second law, the y-component of the force on the particle and thus equals $-\partial V / \partial y$. Similarly, $m\ddot{x}$, the x-component of the force, equals $-\partial V/\partial x$. Now

$$\frac{\partial V}{\partial y} = \frac{dV}{dr}\frac{\partial r}{\partial y} = \frac{y}{r}\frac{dV}{dr},$$

and similarly,

$$\frac{\partial V}{\partial x} = \frac{x}{r}\frac{dV}{dr}.$$

Therefore

$$\frac{dM}{dt} = \left(-\frac{xy}{r} + \frac{yx}{r}\right)\frac{dV}{dr} = 0.$$

Thus the angular momentum does not change with time; it is conserved.

Conservation of electric charge is related to symmetry of the laws of nature under a group of global phase transformations. The description of that symmetry is less straightforward than for the others, and we forgo its description and the derivation of the conservation from it.

7.5. Cosmic Considerations

Let us now return to the analysis of nature into initial states and the laws of nature. That division, whereby the laws of nature and initial states are considered as being independent, seems to be satisfactory for our understanding of nature from the microscopic scale to the macroscopic. It seems to work even for astronomical phenomena, where, for example, we think of different planetary systems or even galaxies as evolving according to the same laws of nature but from different initial conditions. That dichotomy fades away, however, at nature's extreme scales, at the subnuclear scale, the scale of elementary particles, and at the scale of the Universe. A currently widely accepted cosmological scenario holds that the Universe had its beginning in a "big bang" in the far past. Can the evolution of the Universe be analyzed in terms of initial conditions and subsequent development according to the laws of nature? Perhaps, if one considers our Universe to be only one of a variety of possible universes, any one of which could have come into being at the time of the big bang. Then our Universe would be the result of whatever chance initial state initiated its evolution. And the other possible universes? Are they but theoretical possibilities, or did they in some sense all come into being but are inaccessible to us? Such speculations, however intriguing, can never be proved or disproved by science, since we cannot experiment with different initial states for the Universe and follow the evolution of each. That way of looking at things seems to assign to the laws of nature a certain primacy over the Universe itself.

Another point of view, which some people, including myself, find philosophically more satisfactory, is that, since the Universe is, by very definition, the totality of all we have access to, it is meaningless to consider other possible universes, and we must honor the Universe with the distinction of being the only one possible according to the laws of nature. But what are the laws of nature, if not a description of the evolution of the Universe? What that point of view holds is that the big bang, the Universe, the initial state, and the laws of nature are all intimately intertwined, are all aspects of a single, self-consistent situation. Another aspect of that situation is the variety of elementary particles found in nature. While cosmologists are asking why the Universe is as it is on the cosmic scale, other scientists are asking why it is as it is on the subnuclear scale, why nature presents us with the kinds of elementary particles that it does and no other kinds. Is that phenomenon explainable on the basis of the known laws of nature or must new laws of nature be found that relate more directly to the big bang and its immediately subsequent developments, when the present variety of elementary particles presumably came into existence? The latter possibility seems the more likely. That point of view might be summarized by the statement that things are as they are because they cannot be otherwise. See [M49].

How is symmetry connected with all that? Every aspect of that self-consistent entity, the Universe, exhibits various symmetries and approximate symmetries, and we must try to understand how they, too, fit into the all-encompassing picture. The following tale is possibly a small part of that picture.

For every allowed state of any elementary particle having nonzero mass the spatially inverted (i.e., transformed by point inversion) state is also allowed by nature. And the photon, although it is massless, also has that property. Now, the law of nature called the strong interaction, which affects certain of the massive particles (but not all), possesses space-inversion symmetry. And the law of nature called the electromagnetic interaction, which involves the photon and all electrically charged elementary particles, all of which are massive, is also inversion symmetric. Thus we have the situation where certain laws of nature and the entities whose behavior they govern possess the same symmetry, space-inversion symmetry, also called P symmetry, where P (for parity) denotes the point-inversion transformation. In addition, the transformation of particle–antiparticle conjugation, denoted C, is a symmetry both of the strong and electromagnetic interactions and of the massive elementary particles and the photon; for every kind of such particle there is a kind that is the antiparticle of it (with certain ones, including the photon, being their own antiparticles), and all states allowed for particles are also allowed for their antiparticles and vice versa. So again we have the same symmetry, C symmetry, for certain laws of nature and the entities they apply to.

On the other hand, the neutrinos, apparently massless like the photon (although the possibility of their possessing a small mass has not been ruled out), seem to be deficient in those symmetries. The space-inversion image of an allowed neutrino state is not a state allowed by nature. It would be, if the image particle were an antineutrino, for the allowed states of antineutrinos are just the forbidden states of neutrinos. And the inverted states of antineutrinos are forbidden to antineutrinos but allowed for neutrinos. It is as if only left-handed neutrinos and righ-handed antineutrinos are permitted (or, perhaps better, required) to exist by nature, while the existence of right-handed neutrinos and left-handed antineutrinos is excluded. Thus neither P nor C is a symmetry of neutrinos, while the combined transformation of point inversion and particle–antiparticle conjugation, denoted CP, is a symmetry transformation. Now the strong and electromagnetic interactions do not affect neutrinos, while the law of nature called the weak interaction does. (It also affects other particles.) The weak interaction possesses neither P symmetry nor C symmetry; it does possess CP symmetry.

Are the P and C symmetries of the objects upon which the strong and electromagnetic interactions act basically caused by the P and C symmetries of the interactions themselves, or vice versa, or are the interactions and the elementary particles, along with their symmetries, parts of a self-consistent situation with neither more basic than the other? And are the P and C asymmetries and CP symmetry of the neutrinos fundamentally a result of the P and C asymmetries and CP symmetry of the weak interaction, or vice versa, or is it all a self-consistent whole? And how does all that relate to conditions prevailing at the earliest stages of the evolution of the Universe, when the kinds of elementary particles that we know today presumably came into being? Why were the other-handed versions of the neutrinos and antineutrinos not created then? Did the laws of nature even then preclude their existence? Or did the chance creation of the first neutrino–antineutrino pair as a left-handed neutrino and a right-handed antineutrino cause the as yet "plastic" laws of nature to "crystallize" into such a form that would from that moment onward allow proliferation of the existing versions of neutrino and antineutrino and forbid the existence of the other-handed versions (spontaneous symmetry breaking)? Or should we avoid the question by restructuring our concepts to make CP the only meaningful transformation, so that there would be no reason to consider the hypothetical possibility of other versions of neutrinos and antineutrinos? But then why are P and C separately relevant for the strong and electromagnetic interactions? See [M31].

Our pursuit of symmetry and of the understanding of symmetry carries us to the very frontiers of modern science and even to speculations out beyond.

7.6. Summary of Chapter Seven

Taking the active view of transformations in Section 7.1, we formalized the notion of symmetry of the laws of nature, or symmetry of evolution. A symmetry transformation of the laws of nature is a transformation that commutes with the evolution transformation. We saw that the fundamental point underlying the formalism is that a symmetry of evolution is an indifference of nature, whereby the laws of nature ignore some aspect of states of physical systems. We also considered the meaning of a time-reversed process and time-reversal symmetry of the laws of nature.

In Section 7.2 we saw that symmetry of the laws of nature determines an equivalence relation in the state space of a system, the equivalence of states that are indistinguishable by the laws of nature. Applying the equivalence principle, developed in Section 5.2, we obtained the equivalence principle for processes in quasi-isolated systems: Equivalent states, as initial states, *must* evolve into equivalent states, as final states, while inequivalent states *may* evolve into equivalent states. That led to the symmetry principle for processes in quasi-isolated systems: The "initial" symmetry group (that of the cause) is a subgroup of the "final" symmetry group (that of the effect). And that in turn led to the general symmetry evolution principle: For a quasi-isolated physical system the degree of symmetry cannot decrease as the system evolves, but either remains constant or increases. The latter principle is so general as to be quite useless.

So rather than consider the symmetry group of all of state space, we looked in Section 7.3 at the symmetry group of only a single state, the group of permutations of the equivalence subspace (equivalence class) to which it belongs. With the assumption of nonconvergent evolution, that led to the special symmetry evolution principle: The degree of symmetry of the state of a quasi-isolated system cannot decrease during evolution, but either remains constant or increases. Or equivalently: As a quasi-isolated system evolves, the populations of the equivalence subspaces (equivalence classes) of the sequence of states through which it passes cannot decrease, but either remain constant or increase.

For systems possessing nonconvergent evolution of microstates and convergent evolution of macrostates to macrostates of stable equilibrium, such as a gas, the special symmetry evolution principle gave the theorem that the degree of symmetry of a macrostate of stable equilibrium must be relatively high.

The special symmetry evolution principle and the second law of thermodynamics are very similar. That suggested a functional relation between entropy and degree of symmetry, which in fact is already known.

In Section 7.4 we mentioned the intimate and fundamental relation between conservation and symmetry of the laws of nature and saw how that comes about in very simple systems for: conservation of energy and temporal-displacement symmetry, conservation of linear momentum and spatial-displacement symmetry, and conservation of angular momentum and rotation symmetry.

We indulged in considerations of symmetry and asymmetry of the Universe, of the laws on nature, and of the elementary particles in Section 7.5.

CHAPTER EIGHT

Symmetry: The Concept

In this chapter we start our conceptual study of the general theory of symmetry in science, which is continued in the following two chapters. The conceptual approach to symmetry, in contrast to the symmetry formalism developed in Chapters 4–7, is the best approach for understanding the concepts involved in symmetry and the significance of symmetry in science.

First we examine the concept of symmetry at its most general. We see just what the essence of symmetry, as well as of asymmetry and approximate symmetry, is. It is shown how they involve change and lack of change and how a change involves a frame of reference. That leads to consideration of symmetry of the Universe. Finally, we learn how analogy is a symmetry.

The approach is conceptual. No formalism is used, and no special mathematics is needed beyond some geometry. For the problems a slight acquaintance with some notions of nuclear physics would be helpful.

This chapter, and after it Chapters 9 and 10, can be read immediately following Chapter 1, if you choose the "concept" track. After Chapter 10 you can return to Chapter 2 to continue the study of the general theory of the application of symmetry in science. If the "application" track is chosen, on the other hand, and you continue after Chapter 1 to Chapter 2 and subsequent chapters, reaching the present chapter only after Chapter 7, then Chapters 8–10 will fill in the conceptual foundations of symmetry in science.

8.1. The Essence of Symmetry

Our discussion of what the essence of symmetry is actually started in Chapter 1. You are advised to review that chapter at this point. There we state that what symmetry boils down to in the final analysis is that the situation possesses the possibility of a change that leaves some aspect of the situation unchanged. A few simple examples are presented. We then present the concisely formulated conceptual definition of symmetry:

Symmetry is immunity to a possible change.

157

Accordingly, when we do have a situation for which it is possible to make a change under which some aspect of the situation remains unchanged, i.e., is immune to the change, then the situation can be said to be *symmetric under the change with respect to that aspect*.

We point out the two essential components of symmetry:

1. *Possibility of a change.* It must be possible to perform a change, although the change does not actually have to be performed.

2. *Immunity.* Some aspect of the situation would remain unchanged, if the change were performed.

If a change is possible but some aspect of the situation is not immune to it, we have *asymmetry*. Then the situation can be said to be *asymmetric under the change with respect to that aspect*.

If there is no possibility of a change, then the very concepts of symmetry and asymmetry will be inapplicable. For example, if the property of color is not an ingredient of the specification of a plane figure, then the change of, say, color interchange will not be a possible change for such a figure. Thus color interchange symmetry or asymmetry will not be conceptually applicable to the situation. Or alternatively, one might say that such a plane figure will possess trivial symmetry under such a change. One might say that all its aspects will be trivially immune to such a change. It is a matter of taste, but I tend to prefer calling it inapplicability rather than triviality.

If, however, color *is* included in the specification of a figure, then color interchange will become a possible change for it. For example, if the figure is black and white, it will be symmetric under red–green interchange with respect to appearance. Interchange red and green, and nothing will happen to the figure. If the figure is black and green, it will be asymmetric under the same change with respect to the same aspect. Interchange red and green, and the figure will become black and red, which is not the same as black and green.

Approximate symmetry is approximate immunity to a possible change. There is no approximation in the change or in its possibility; it must indeed be possible to perform a change. The approximation is in the immunity. Some aspect of the situation must change by only a little, however that is evaluated, when some change is performed. Then the situation can be said to be *approximately symmetric under the change with respect to that aspect*.

For example, the figure of Fig. 8.1 possesses approximate twofold rotation symmetry with respect to its appearance. Under 180° rotation about its center its appearance changes, but only by a little. And the bilateral symmetry of humans and other animals is in reality also only approximate. Not only do the internal organs not all possess that symmetry, but even for external appearance the symmetry is never exact. For instance, the fingerprints of one hand are not the mirror images of the corresponding fingerprints of the other hand, and the hand and foot of one side (usually the right side for right-handed people) are almost always slightly longer than those of the other side.

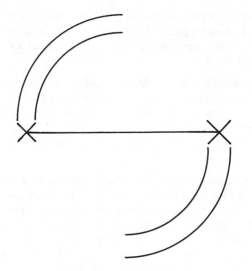

Fig. 8.1. Approximate twofold rotation symmetry.

Approximate symmetry is a softening of the hard symmetry–asymmetry dichotomy. The extent of deviation from exact symmetry that can still be considered approximate symmetry will depend on the context and the application and could very well be a matter of personal taste. The same figure, for example that of Fig. 8.1, might be considered approximately symmetric (or slightly asymmetric) by some observers, such as myself, while others might consider it very asymmetric (or nowhere near symmetric).

PROBLEMS

1. Consider the black–white figure of Fig. 8.2 and various changes, including compound changes, that are possible for it. With respect to which of its aspects, if any, is the figure symmetric or asymmetric under each of those changes? Now imagine that the figure is but part of an infinite pattern and answer the same question for the infinite pattern.

Fig. 8.2. Black–white figure for problem.

2. Consider the symmetry or asymmetry of the line "Beauty is truth, truth beauty," from John Keats's (English poet, 1795–1821) *Ode to a Grecian Urn*, under beauty–truth interchange with respect to meaning, meter, graphical appearance, and other aspects you might think of.

3. What are the rotation, mirror-reflection, and point-inversion symmetries and asymmetries of a regular tetrahedron with respect to its position and orientation in space? What are the spatial-displacement, rotation, mirror-reflection, and point-inversion symmetries and asymmetries of the diamond lattice with respect to the same aspects? (Models might help. In case the transformations are not familiar, see Section 4.3.)

4. What are the rotation, mirror-reflection, and point-inversion symmetries and asymmetries with respect to position and orientation in space of the other four Platonic solids: the cube, the regular octahedron, the regular dodecahedron, and the regular icosahedron? (Models might help. See Section 4.3, if the transformations are not familiar.) Compare your results for the cube with those for the octahedron and similarly for the dodecahedron and the icosahedron. They should be the same for the members of each pair. What is the geometric significance of that? In what way is the regular tetrahedron an exception?

8.2. How Is Change?

Change is the bringing about of something different. For a difference to exist, in the sense of having physical meaning, a physical gauge for the difference, a *frame of reference*, is needed. Thus the existence of a frame of reference is necessary to give existence to the difference and to the possibility of change. And the nonexistence of an appropriate frame of reference makes a supposed change impossible.

In order to be capable of gauging a difference, a frame of reference cannot be immune to the change that brings about that difference. It cannot be immune to the change for which it is intended to serve as reference. Otherwise, it could not serve its purpose.

For example, the change of spatial displacement brings about a difference in location. A set of tape measures as coordinate axes could serve as a frame of reference for that, but not if the tapes are themselves immune to displacement, i.e., not if the tapes are infinite and homogeneous (unmarked). Marked axes, which can indeed gauge differences in location and can thus serve as a frame of reference for displacements, are themselves affected by displacements.

A frame of reference is a changeable aspect of a situation. Now, any changeable aspect of a situation can serve as a frame of reference for that change in the situation, since it is tautological that a changeable aspect of a situation is not immune to its own change. A changeable aspect of a situation allows the possibility of a change. Indeed, we can say that it represents the possibility of a change and that any possibility of a change is represented by a changeable aspect of the situation. So a situation will possess symmetry if and only if it has *both* an aspect that can change (giving the possibility of a change) *and* an aspect that does not change concomitantly (giving the immunity to the possible change). In other words, the possibility of a change, which is a necessary component of symmetry, is contingent upon the existence of an asymmetry of the situation under the change. And hence the result:

Symmetry implies asymmetry.

Consider, for example, the equilateral triangle of Fig. 1.2. Its appearance is an aspect of it that is immune to 120° rotation about its center. Its microscopic, molecular structure happens to be an aspect of it that is not immune to 120° rotation. But even if the triangle were microscopically perfectly homogeneous and isotropic and were thus immune to 120° rotation in all its aspects, or even if the triangle were considered to be a nonstructured abstract figure, the triangle is not a universe in itself. The *total* situation, that of the equilateral triangle together with its surroundings, does possess aspects that are not immune to 120° rotation and that can thereby serve as frame of reference for 120° rotation. The walls of the room could serve as frame of reference, for instance, since they are asymmetric under 120° rotation. Thus rotation by 120° is indeed a change. The equilateral triangle is symmetric in the context of its surroundings. It is symmetric under 120° rotation thanks to its surroundings' lack of immunity to 120° rotation, thanks to the asymmetry of the total situation, triangle plus surroundings, under 120° rotation.

Our results concerning symmetry, change, immunity, frame of reference, and asymmetry can be summarized by the following diagram, where arrows denote implication:

Symmetry ⟶ ⎡ Possibility → Reference for → Asymmetry under
⎪ of a change   the change   the change
⎣ Immunity to
the change.

Thus, for there to be symmetry, there must concomitantly be asymmetry under the same change that is involved in the symmetry. For every symmetry there is an asymmetry tucked away somewhere in the world.

So symmetry implies asymmetry. That relation is not symmetric, since asymmetry does not imply symmetry, at least not in the same sense that symmetry implies asymmetry, in the sense that actual symmetry implies actual asymmetry, as was demonstrated above. However, asymmetry does imply symmetry in the limited sense that the lack of immunity to a possible change implies the conceptual possibility of immunity to that change. Thus actual asymmetry implies merely the conceptual possibility of, not actual, symmetry.

PROBLEMS

1. State possible frames of reference for each of the changes you considered in Problem 1 of the preceding section.

2. What are possible frames of reference for beauty-truth interchange in Problem 2 of the preceding section?

3. For the changes of Problems 3 and 4 of the preceding section declare what could serve as frames of reference.

4. What frames of reference could allow symmetry, approximate symmetry, or asymmetry of atomic nuclei with respect to nuclear properties under proton-neutron interchange?

8.3. Symmetry of the Universe

Another example of symmetry implying asymmetry: The laws of nature have been found to be the same wherever they have been investigated, and it seems to be consistent with observation to assume that the laws of nature are the same everywhere in the Universe. That is a symmetry of the Universe. The laws of nature we discover through our investigations are immune to the possible change of displacing our laboratory from here to there. Spatial displacement is indeed a change, because here and there are different from each other. Here might be on the surface of the Earth, while there might be inside a star. The inhomogeneous distribution of matter in the Universe serves as a frame of reference for spatial displacement. Thus the Universe possesses the symmetry that the laws of nature are immune to spatial displacement. Spatial displacement is a change thanks to the Universe's asymmetry, that the distribution of matter in it is not the same everywhere, is not immune to spatial displacement.

Let us now imagine a hypothetical universe that might seem to be even more symmetric than the Universe actually is. Imagine a universe in which not only the laws of nature are the same everywhere, but so is the distribution of matter. Indeed, imagine a completely homogeneous universe, i.e., a universe possessing exactly the same properties at every location. Such a universe might seem to be perfectly symmetric under spatial displacement in that *all* its aspects are immune to the possible change of spatial displacement. (That would be in contrast to the actual Universe, which is not perfectly symmetric under spatial displacement, but only with respect to the laws of nature.)

Yet, where in such a universe is the frame of reference for spatial displacement? What makes spatial displacement a change? Nothing at all! For such a universe there is no frame of reference for spatial displacement, so spatial displacement is no change. Such a universe is not more symmetric under spatial displacement than is the real Universe. Indeed, it possesses no spatial displacement symmetry at all. But neither is it asymmetric. The very concepts of spatial displacement symmetry and asymmetry are inapplicable to it. That is because it does not have the possibility of spatial displacement. And that is because it possesses no frame of reference for spatial displacement: there is no differentiation among locations in it; here it no different from there in any respect.

I have noticed that some people, when trying to imagine a perfectly homogeneous universe, do not go far enough and imagine themselves as an observer who can flit about from place to place within the universe, making observations and labeling positions here and there. That kind of observer is an inhomogeneity and is absolutely excluded from a homogeneous universe. If an observer is to be imagined at all, he or she must be homogeneous as well, "smeared out" smoothly and evenly over all space. Such an observer could not distinguish one location from another.

Another point I would like to mention in this connection, although it is not directly relevant to our discussion, is that since all locations in a hypothetical homogeneous universe are completely undifferentiated, all locations are identical and are actually one and the same location. Thus such a universe lacks spatiality altogether. (A homogeneous space as a *mathematical model* of the Universe is another matter altogether. There we conceptually impose spatial differentiation and a frame of reference, i.e., a coordinate system, on it.)

Since the Universe is everything and an imagined universe is imagined to be everything, no external frame of reference can be imposed on it. If the Universe, or a universe, does not contain a frame of reference within itself, then such a frame of reference is meaningless. That is in contrast to the equilateral triangle of Fig. 1.2. Even if the triangle were perfectly homogeneous and isotropic and had no frame of reference of its own for rotation by 120°, such a frame of reference would be imposed on it by its surroundings, say, by the walls. The Universe, or a universe, has no surroundings. If something makes no difference to the Universe, then there is nothing else for it to make a difference to.

Hence the cosmic conclusion:

The Universe cannot possess perfect symmetry.

Perfect symmetry of the Universe under some change would mean that there exists the possibility of that change for the Universe and that *all* aspects of the Universe are immune to it, i.e., that there exists no aspect of the Universe that is not immune to the change. But then the Universe would possess no frame of reference for the change. So the change would not be possible for it, and there would be no symmetry. The above example of putative perfect spatial displacement symmetry of a universe shows how that works in a specific case.

Indeed, perfect symmetry of the Universe is a contradiction in terms. Perfect symmetry of the Universe would mean that all aspects of the Universe are immune to some change, that there is no aspect of the Universe that is not immune to the change, thus no asymmetry under the change. That is a contradiction, since perfect symmetry of the Universe would not be fulfilling a necessary condition for it to be symmetry at all, which is that there be asymmetry under the same change. The Universe would possess no frame of reference for the change. So the change would not be possible for the Universe, and there would be no symmetry. Therefore, again, the Universe cannot possess perfect symmetry.

PROBLEMS

1. Try to imagine a universe possessing perfect symmetry under proton–neutron interchange and assess the implications thereof. For instance, what would the terms "proton" and "neutron" designate in such a universe?

2. What about a perfectly mirror-reflection symmetric universe (for all possible mirrors)?

8.4. Analogy as Symmetry

A very important kind of symmetry, which is not often though of as a symmetry, is *analogy*. Analogy is the immunity of the validity of a relation or statement to changes of the elements involved in it.

To see what we actually have here, consider, for example, the statement, "An animal has a relatively long tail." That statement involves a single element, an animal. But the element is not unique, since there are more than just a single animal in the world. Indeed, one can say, for example, "This deer has a relatively long tail," or "This squirrel has a relatively long tail." But the statement is not valid for all animals. Deer do not have relatively long tails, while squirrels do. Nevertheless, there are, in fact, more than just a single animal for which the statement is valid. Squirrel *A* has a relatively long tail, squirrel *B* also has a relatively long tail, squirrel *C* does too, so does squirrel *D*, and so on for all squirrels as well as for certain other animals. That statement is an expression of analogy among animals: All relatively long-tailed animals are analogous in that, whatever their differences, they all possess the common property of having a relatively long tail. And as a fringe benefit we have that all relatively non-long-tailed (i.e., medium-, short-, and no-tailed) animals are analogous in that, whatever their differences, they all possess the common property that their tails are not relatively long.

That this analogy is a symmetry is seen by nothing: (1) There is the possibility of a change. Since the statement is applicable to more than a single animal, the animal to which it is applied can be switched. (2) An aspect of the statement, its validity, is immune to such a change. The statement is valid just as well for squirrel *A*, for squirrel *B*, for *C*, etc., who each proudly waves a relatively long tail.

For another, similar analogy consider the statement, "An astronomical body moves along an elliptical orbit with the Sun at one of its foci." That statement too involves a single element that is not unique; there are more than one astronomical body in the cosmos. For example, one can say, "The Moon moves …," or "Venus moves. …" But the statement is not valid for all astronomical bodies. The Moon does not move in that way, while Venus does. There are, however, more than a single astronomical body for which the statement is valid. They include all the planets of the solar system, for whom the statement becomes Kepler's first law of planetary motion (Johannes Kepler, German astronomer and mathematician, 1571–1630), as well as the asteroids and some of the comets. That statement is an expression of analogy among astronomical bodies: All those bodies that move along elliptical orbits with the Sun at one of their foci are analogous in that, whatever their differences, they all move in just that way. And we also have that all the other astronomical bodies, such as stars and moons, are analogous in that, whatever their differences, they all do not move along ellipses with the Sun at one of the foci.

That analogy too is a symmetry: (1) There is the possibility of a change. Since the statement is applicable to more than a single astronomical body, the body to which it is applied can be switched. (2) An aspect of the statement, its validity, is immune to such a change. The statement is valid just as well for Pluto, for Neptune, for Uranus, etc., each of which moves along an elliptical orbit with the Sun at one of its foci.

Now consider a statement involving a pair of elements, "X is the locus of all points equidistant from a given point (its center), all points lying in Y." This statement invovles a pair of elements, (X, Y), where X and Y can be any geometric objects and are certainly not unique. For instance, one can say, "A triangle is the locus … lying in an ellipsoid," or "A circle is the locus … lying in a plane." But the statement is not valid for all pairs of geometric objects. It is not true for the pair (a triangle, an ellipsoid), while it does hold for the pair (a circle, a plane). There are more than one pair (X, Y) for which it is valid. Three of them are: (a pair of points, a line), (a circle, a plane), and (a spherical surface, space). That statement of a relation between X and Y is an expression of analogy among pairs of geometric objects: All those pairs (X, Y) whose elements X and Y fulfill the relation as stated are analogous in that, whatever their differences, they all fulfill the relation. And in addition, all those pairs whose elements do not fulfill the relation are analogous in that, whatever their differences, they do not fulfill the relation.

Analogies involving pairs of elements are often put in the form: A is to B as C is to D as. … For the present example this form is: A pair of points is to a line as a circle is to a plane as a spherical surface is to space. … (That form of expressing an analogy can easily be generalized for relations involving any number of elements.) The symmetry here is that:

(1) Since the statement is applicable to more than one pair of geometric objects, the pair to which it is applied can be switched.
(2) The validity of the statement is immune to such a change.

Now, consider the experimental setup of a given sphere rolling down a fixed inclined plane, with the experimental procedure of releasing the sphere from rest, letting it roll for any time t, and noting the distance d the sphere rolls in that time. Performing n such experiments, we collect n data pairs (or data points) $(t_1, d_1), \ldots, (t_n, d_n)$. Those pairs obey the relation $d_k = bt_k^2$, where b is a positive proportionality constant, for $k = 1, \ldots, n$. That is a relation involving two elements, as in the preceding example. Then n data pairs, as well as an infinity of potential data pairs, are analogous in that they all obey the same relation, $d = bt^2$, and in that sense t_1 is to d_1 as t_2 is to d_2 as.... All other (t, d) pairs, which do not obey the relation $d = bt^2$, are also analogous in that they do not obey the relation. The symmetry is that we can switch among actual and potential data pairs, and however we switch among them, the relation between t and d remains the same.

Additional physics analogies can be found in [M57].

With the help of the four examples above we now see how analogy, as the immunity of the validity of a relation or statement under changes of the elements involved in it, is indeed what we thought we understood by the term "analogy" before we found ourselves hopelessly confused by such a weird definition. The reason for such a definition of analogy, besides its being a good one, is that it directly exposes the symmetry that is analogy, since it implies:

(1) the possibility of a change, the change of elements involved in the relation or statement; and
(2) the immunity of the validity of the relation or statement to such a change.

Note that analogy implies and is implied by *classification*. An analogy imposes a classification by decomposing the set of elements or element pairs, triples, etc. to which the statement or relation is applicable into classes of analogous elements or pairs, triples, etc. For example, among animals the statement, "An animal has a relatively long tail," separates all animals into a class of relatively long-tailed animals, those animals for whom the statement is valid, and a class of relatively non-long-tailed animals, those for whom the statement is false. In the astronomical example the statement, "An astronomical body moves along an elliptical orbit with the Sun at one of its foci," decomposes all astronomical bodies into a class of those for which the statement is valid, the most notable of which are the planets of the solar system, and a class of astronomical bodies that do not move according to the statement, which includes the planetary moons and all the stars, among others.

In the geometric example the relation, "X is the locus of all points equidistant from a given point (its center), all points lying in Y," separates all pairs of geometric objects into a class of those pairs for which the relation holds, the best known of which are (a pair of points, a line), (a circle, a plane), and (a spherical surface, space), and a class of those that do not fulfill the relation, such as (a triangle, an ellipsoid) and (a hyperboloid, space). And in the laboratory example the relation $d = bt^2$ decomposes all (t, d) pairs into a class of those obeying the relation, i.e., all actual and potential data pairs for the experiment, and a class of those for which $d \neq bt^2$, those that cannot be data for the experiment.

Conversely, a classification defines the analogy of belonging to the same class. If any set is decomposed into mutually exclusive classes, then the very property of belonging to the same class will define an analogy among the elements of the set. For instance, the kids in a school can be, and for administrative purposes are, classified by grade. That makes all pupils in the same grade analogous. Or, motor vehicles can be classified by the number of axles. That classification makes all vehicles with the same number of axles analogous, which might find expression in the toll rate on toll roads.

PROBLEMS

1. There are very many systems—mechanical, electromagnetic, chemical, biological, whatever—that exhibit the following behavior: When the system is in a state of stable equilibrium and some parameter of the system is, by external intervention, changed slightly from its equilibrium value and then left to vary freely, the parameter will perform damped harmonic oscillations. A parameter might be a position, a pressure, a voltage, a concentration, or a temperature, for example. All such systems are thereby analogous. State the analogy in the form of the definition presented in this section. Show that the analogy is a symmetry. Describe the classification implied by the analogy.

2. People can be classified by their blood type (A, B, AB, or O) and their Rh group (+ or −). Describe the analogy implied by the classification and state it in the form of the definition presented in this section. Show that it is a symmetry.

3. Find a useful analogy, or equivalently a classification, among the chemical elements (you do not have to be original). State it in the form of the definition presented in this section and show that it is a symmetry.

4. The systematic classification of plants and animals into species, genus, family, and so on implies an analogy among living beings. Describe that analogy. Express it as a symmetry.

8.5. Summary of Chapter Eight

In Section 8.1 (and Chapter 1) we saw that the essence of symmetry is that a situation possesses the possibility of a change that leaves some aspect of the situation unchanged, which is formulated concisely as: Symmetry is immunity to a possible change. Asymmetry is the possibility of a change under which some aspect of the situation changes concomitantly. If there is no possibility of a change, then the concepts of symmetry and asymmetry will be inapplicable. Approximate symmetry is approximate immunity to a possible change.

In Section 8.2 we considered the significance of change and saw that a frame of reference is necessary for the possibility of change, where a frame of reference is any changeable aspect of the situation. Thus symmetry implies asymmetry; a situation will possess symmetry if and only if it has *both* an aspect that can change (giving the possibility of a change) *and* an aspect that does not change concomitantly (giving the immunity to the possible change). It then follows, as we saw in Section 8.3, that perfect symmetry of the Universe is impossible.

In the discussion of Section 8.4 we saw that analogy, which is the immunity of the validity of a relation or statement to changes of the elements invovled in it, is a kind of symmetry. We also saw that analogy implies and is implied by classification.

CHAPTER NINE

Symmetry in Science

In this chapter we briefly review what science is about, and we see that it strongly involves reduction. It is shown that reduction is a symmetry. We consider three ways reduction is used in science: observer and observed, quasi-isolated system and surroundings, and initial state and evolution. Reproducibility, predictability, and reduction compose the triple foundation of science. The former two are shown to be symmetries and to imply analogies. Thus symmetry is of major importance in the foundation of science.

As in the preceding chapter the discussion is conceptual, and the physical significance of the concepts is emphasized throughout. There is no formalism, and no special mathematical preparation is needed.

9.1. Science

For the purpose of our discussion we take this definition of *nature*:

> *Nature is the material universe with which*
> *we can, or can conceivably, interact.*

The *material universe* is everything having a purely material character. To *interact* with something is to act upon it and be acted upon by it. That implies the possibility of performing observations and measurements on it and of receiving data from it, which is what we are actually interested in. To be able *conceivably* to interact means that, although we might not be able to interact at present, interaction is not precluded by any principle known to us and is considered attainable through further technological research and development. Thus nature, as the material universe with which we can, or can conceivably, interact, is everything of purely material character that we can, or can conceivably, observe and measure.

We live in nature, observe it, and are intrigued. We try to understand nature in order both to improve our lives by better satisfying our material needs and desires and to satisfy our curiosity. And what we observe in nature is a complex of phenomena, including ourselves, where we are related to all of nature, as is implied by our definition of nature as the material universe with which *we can, or can conceivably, interact*. The possibility of interaction is what relates us to all of nature and, due to the mutuality of interaction and of the relation it brings about, relates all of nature to us. It then follows that all aspects and phenomena of nature are actually interrelated, whether they appear to be so or not; whether they are interrelated independently of us or not, they are certainly interrelated through our mediation. Thus all of nature, including *Homo sapiens*, is interrelated and integrated.

Now we come to *science*:

> *Science is our attempt to understand the*
> *reproducible and predictable aspects of nature.*

And, I repeat, nature is the material universe with which we can, or can conceivably, interact. By *our* we mean that science is a human endeavor and is shaped by our modes of perception and our mental makeup. *Attempt* means that we try but might not always succeed. By *understand* we mean be able to explain. We do that by looking for *order* among the reproducible and predictable aspects of nature, formulating *laws*, and devising *theories*. See [M52].

Reproducibility means that experiments can be replicated in other laboratories and in the same laboratory, thus making science a common endeavor that is as objective as possible. (Reproducibility is treated in more detail in Section 9.6.) *Predictability* means that among the phenomena investigated, order can be found, from which laws can be formulated, predicting the results of new experiments. (See Section 9.7 for a more detailed treatment of predictability.) There is no claim here that all of nature's aspects are reproducible and predictable. Indeed, they are not. For example, according to our understanding of nature's quantum aspect, individual submicroscopic events, such as the radioactive decay of an unstable nucleus, are inherently unpredictable. (However, the statistical properties of many submicroscopic events, such as the half-life of a radioactive isotope, may be predictable.) And perhaps the behavior of an individual organism is inherently not completely predictable either. But all such aspects of nature lie outside the domain of concern of science. Reproducibility and predictability form the dual foundation upon which science firmly rests. If either is lacking, science will be unable to operate.

PROBLEM

As science attempts to comprehend larger- and larger-size phenomena of nature, actual reproducibility is replaced by declared reproducibility, in the sense that, even if we cannot actually replicate the effect, such as a volcano or the birth of a star, nature supplies us with a sufficient quantity and variety. But as the size increases to truly gigantic, such as superclusters of galaxies, that reasoning becomes tenuous. Moreover, when the Universe is considered as a whole, we cannot even declare reproducibility. Whatever metaphysical ideas one might have about universes, science deals solely with the single universe we are part of, with the Universe, which is thus irreproducible by any meaning of the word. After that lengthy introduction, consider and discuss to what extent cosmology (the study of the working of the cosmos, the Universe as a whole) and cosmogony (considerations of the origin of the cosmos) can be thought of as branches of science.

9.2. Reduction as Symmetry

But how are we to grasp the wholeness, the integrality, that is nature? When we approach nature in its completeness, it appears so awesomely complicated, due to the interrelation of all its aspects and phenomena, that it might seem utterly beyond hope to understand anything about it at all. True, some obvious simplicity stands out, such as day–night periodicity, the annual cycle of the seasons, and the fact that fire consumes. And subtler simplicity can be discerned, such as the term of pregnancy, the relation between clouds and rain, and that between the tide and the phase of the moon. Yet, on the whole, complexity seems to be the norm, and even simplicity, when considered in more detail, reveals wealths of complexity. But, again due to nature's unity, any attempt to analyze nature into simpler component parts cannot but leave something out of the picture.

Holism is the world view that nature can be understood only in its wholeness or not at all. And that includes human beings as part of nature. As long as nature is not yet understood, there is no reason *a priori* to consider any aspect or phenomenon of it as being intrinsically more or less important than any other. Thus it is not meaningful to pick out some part of nature as being more "worthy" of investigation than other parts. Neither is it meaningful, according to the holist position, to investigate an aspect or phenomenon of nature as if it were isolated from the rest of nature. The result of such an effort would not reflect the normal behavior of that aspect or phenomenon, since in reality it is not isolated at all, but is interrelated with and integrated in all of nature, including ourselves.

On the other hand lies the world view called *reductionism*, which holds that nature is indeed understandable as the sum of its parts. According to the reductionist position nature should be studied by analysis, should be "chopped up" (mostly conceptually, of course) into simpler component parts that can be individually understood. (By "parts" we do not necessarily mean actual material parts; the term might be used metaphorically. An example of that is presented in Section 9.5.) A successful analysis should then be followed by synthesis, whereby the understanding of the parts is used to help attain understanding of larger parts compounded of understood parts. If necessary, that should then be followed by further synthesis, by the consideration of even larger parts, compounded of the understood compound parts, and attaining understanding of the former with the help of the understanding achieved thus far. And so on to the understanding of ever larger parts, until we hopefully reach an understanding of all of nature.

Now, each of the poles of holism and reductionism has a valid point to make. Nature is certainly interrelated and integrated, at least in principle, and we should not lose sight of that fact. But if we hold fast to extreme holism, everything will seem so fearsomely complicated that it is doubtful we will be able to do much science. Separating nature into parts seems to be the only way to search for simplicity within nature's complexity. Thus the method of *reduction* might well be thought of as sharing with reproducibility and predictability their eminence as components of the foundation of science, and we might well think of all three as composing the triple foundation of science.

Yet a position of extreme reductionism might also not allow much science progress, since nature might not be as amenable to reduction as this position claims, and the reductionist method of science might eventually run up against a holistic barrier. So science is forced to the pragmatic mode of operating *as if* reductionism were valid and adhering to that assumption for as long as it works. In the meanwhile science continues to operate very well. But it should be kept in mind that the inherent integrity of nature can raise its head at any time and indeed does so. The most well-known aspect of nature's irreducibility is nature's quantum character. See [M7] and [M9].

Reduction in science, the separation of nature into parts that can be individually understood, implies symmetry. The point is that if a reduction separates out a part that can be understood individually, then that part exhibits order and law regardless of what is going on in the rest of nature. In other words, that part possesses aspects that are immune to possible changes in the rest of nature. And that is symmetry. It is then possible to make a change (in the rest of nature) that leaves some aspect of the situation (some aspect of that part of nature) unchanged.

Reduction of nature can be carried out in many different ways. As the old saying goes, there's more than one way to slice a salami. We now consider three ways reduction is commonly applied in science, three ways nature is commonly "sliced up," and we examine the symmetry implied by each.

PROBLEM

Since all of nature, including ourselves, is interrelated and integrated, what conditions must nature fulfill to enable the method of reduction, and thus science, to work?

9.3. Observer and Observed

The most common way of reducing nature is to separate it into two parts: the *observer*—us—and the *observed*—the rest of nature. That reduction is so obvious, it is often overlooked. It is so obvious because in doing science we *must* observe nature to find out what is going on and what needs to be understood. Now, what is happening is this: Observation is interaction, so we and the rest of nature are in interaction, are interrelated, as was pointed out in the preceding section. Thus anything we observe inherently involves ourselves too. The full phenomenon is thus at least as complicated as *Homo sapiens*. Every observation must include the reception of information by our senses, its transmission to our brain, its processing there, its becoming part of our awareness, its comprehension by our consciousness, etc. We appear to ourselves to be so frightfully complicated, that we should then renounce all hope of understanding anything at all.

So we reduce nature to us, on the one hand, and the rest of nature, on the other. The rest of nature, as complicated as it might be, is much less complicated than all of nature, since we have been taken out of the picture. We then concentrate on attempting to understand the rest of nature. (We also might, and indeed do, try to understand ourselves. But that is another story.) However, as we saw above, since nature with us is not the same as nature without us, what right have we to think that any understanding we achieve by our observations is at all relevant to what is going on in nature when we are not observing? The answer is that in principle we simply have no such right *a priori*. What we are doing is *assuming*, or adopting the working hypothesis, that the effect of our observations on what we observe is sufficiently weak or can be made so, that what we actually observe well reflects what would occur without our observation, and that the understanding we reach under that assumption is well relevant to the actual situation. That assumption might be a good one or it might not, its suitability possibly depending on the aspect of nature that is being investigated. It is ultimately assessed by its success or failure in allowing us to understand nature.

As is well known, the observer–observed analysis of nature is very successful in many realms of science. One example is Newton's explanation (Isaac Newton, English philosopher and mathematician, 1642–1727) of Kepler's laws of planetary motion. That excellent understanding of an aspect of nature was achieved under the assumption that observation of the planets does not affect their motion substantially. In general, the reduction of nature to observer and observed seems to work very well from astronomical phenomena down through everyday-size phenomena and on down in size to microscopic phenomena. However, at the microscopic level, such as in the biological investigation of individual cells, extraordinary effort must be invested to achieve a good separation. The ever-present danger of the observation's distorting the observed phenomena, so that the observed behavior does not well reflect the behavior that would occur without observation, must be constantly circumvented.

At the molecular, atomic, and nuclear levels and at the subnuclear level, that of the so-called elementary particles and their structure, the observer–observed analysis of nature does not work. Here it is not merely a matter of lack of ingenuity or insufficient technical proficiency in designing devices that minimize the effect of the observation on the observed phenomena. Here it seems that the observer–observed interrelation cannot be disentangled *in principle*, that nature holistically absolutely forbids our separating ourselves from the rest of itself. Quantum theory successfully deals with such matters [M9, M7]. From it we learn that nature's observer–observed disentanglement veto is actually valid for *all* phenomena of *all* sizes. Nevertheless, the *amount* of residual observer–observed involvement, after all efforts have been made to separate, can be more or less characterized by something like atom size. Thus an atom size discrepancy in the observation of a planet, a house, or even a cell is negligible, while such a discrepancy in the observation of an atom or an elementary particle is of cardinal significance.

One aspect of the symmetry implied by the observer–observed reduction, when the latter is valid, is that the behavior of the rest of nature (i.e., nature without us) is unaffected by and independent of our observing and measuring. That behavior is thus an aspect of nature that is immune to certain possible changes, the changes being changes in our observational activities. It is just that symmetry that allows the compilation of objective, observer-independent data about nature that is a *sine qua non* for the very existence of science. It is intimately related to the symmetry that is reproducibility, which is discussed in Section 9.6.

Inversely, another aspect of that symmetry is that our observational activity is unaffected by and independent of the behavior of the rest of nature, at least in certain respects and to a certain degree. For example, if we had an ideal thermometer, we would apply exactly the same temperature measurement procedure regardless of the system whose temperature is being taken. (In practice, of course, things are not so simple.) The symmetry here is that our observational activity is an aspect of nature that is (at least ideally) immune to changes in what is being observed. That symmetry, as limited as it might be in practice, allows the setting up of measurement standards and thus allows the meaningful comparison of observational results for different systems. For instance, we can meaningfully compare the temperature of the sea with that of the atmosphere.

PROBLEM

There are fields of study, such as psychology, sociology, and anthropology, in which the observer–observed separation is very problematic, if not altogether impossible. Does that mean those studies are not based on objective, observer-independent data? So to what extent should those fields be considered branches of science? (What difference does it make, and to whom, whether they are so considered or not?)

9.4. Quasi-Isolated System and Surroundings

Whenever we reduce nature to observer and the rest of nature, we achieve simplification of what is being observed, because instead of observing all of nature, we are then observing only what is left of nature after we ourselves are removed from the picture. Yet even the rest of nature is frightfully complicated. That might be overcome by further slicing of nature, by separating out from the rest of nature just that aspect or phenomenon that especially interests us. For example, in order to study liver cells we might remove a cell from a liver and examine it under a microscope.

But what right have we have to think that by separating out a part of nature and confining our investigation to it, while completely ignoring the rest, we will gain meaningful understanding? We have in principle no right at all *a priori*. Ignoring everything going on outside the object of our investigation will be meaningful if the object of our investigation is not affected by what is going on around it, so that it really does not matter what is going on around it. That will be the case if there is no interaction between it and the rest of nature, i.e., if the object of our investigation is an *isolated system*.

Now, an isolated system is an idealization. By its very definition we cannot interact with, thus we cannot observe, an isolated system, so no such animal can exist in nature, where nature is, we recall, the material universe with which we can, or can conceivably, interact. So we have no choice but to deal with nonisolated systems. The known anti-isolatory factors include the various forces of nature, which can either be effectively screened out or can be attenuated by spatial separation [M8]. Additional anti-isolatory factors involve quantum effects and inertia, which can be neither screened out nor attenuated. Thus even the most nearly perfectly isolated natural system is simply not isolated, and I therefore prefer the term "quasi-isolated system" for a system that is as nearly isolated as possible.

The separation of nature into *quasi-isolated system* and *surroundings* will be a reduction, if, in spite of the system's lack of perfect isolation, there are aspects of the system that are nevertheless unaffected by its surroundings. And the fact of the matter is that the investigation of quasi-isolated systems does yield meaningful understanding, thus proving quasi-isolation to be a reduction of nature. Indeed, science successfully operates and progresses by the double reduction of nature into observer and observed and the observed into quasi-isolated system and its surroundings.

One side of the symmetry implied by this reduction is that those aspects of quasi-isolated systems that are not affected by their surroundings are aspects of nature that are immune to possible changes, the changes being changes in the situation of the surroundings. This symmetry is intimately related to the symmetry that is predictability, which is discussed in Section 9.7. Inversely, due to the mutuality of interaction or of lack of interaction, there are also aspects of the surroundings of quasi-isolated systems that are immune to certain changes in the states of the quasi-isolated systems. That is another side of the symmetry implied by this reduction.

PROBLEM

For complex systems whose components are in strong mutual interaction, such as, perhaps, social and economic systems, the reduction to significantly quasi-isolated relatively simple subsystem and surroundings (the rest of the complex system together with the rest of the world) might be difficult or impossible. How should science attempt to comprehend such systems?

9.5. Initial State and Evolution

The previous two ways of reducing nature—separation into observer and observed and separation into quasi-isolated system and its surroundings— are literal applications of the reductionist position. The present way of reducing is a metaphoric application, or a broadening of the idea of a part of nature. Rather than a separation that can usually be envisioned spatially— observer here, observed there, or quasi-isolated system here, its surroundings around it—the present reduction is a conceptual separation, the separation of natural processes into *initial state* and *evolution*.

Things happen. Events occur. Changes take place. Nature evolves. That is the relentless march of time. The process of nature's evolution is of special interest to scientists, since predictability, one of the cornerstones of science, has to do with telling what will be in the future, what will evolve in time. Nature's evolution is certainly a complicated process. Yet order and law can be found in it, when it is properly sliced. First the observer should separate him- or herself from the rest of nature. Then he or she should narrow the scope of investigation from all of the rest of nature to quasi-isolated systems and investigate the natural evolution of such systems only. Actually, it is only for quasi-isolated systems that order and law are found.

Finally, and this is the present point, the natural evolution of quasi-isolated systems should be analyzed in the following manner. The evolution process of a system should be considered as a sequence of *states* in time, where a state is the condition of the system at any time. For example, the solar system evolves, as the planets revolve around the Sun and the moons revolve around their respective planets. Now imagine that some duration of this evolution is recorded on a reel of photographic film or on a videocassette. Such a recording is actually a sequence of still pictures. Each still picture can be considered to represent a state of the solar system, the positions of the planets and moons at any time. The full recording, the reel or cassette, represents a segment of the evolution process.

Then the state of the system at every time should be considered as an *initial state*, a precursor state, from which the following remainder of the sequence develops, from which the subsequent process evolves. For the solar system, for instance, the positions and velocities of the planets and moons at every single time, such as when it is twelve o'clock noon in Bethesda on 20 October 1992, say, or any other time, should be considered as an initial state from which the subsequent evolution of the solar system follows.

When that is done, when natural evolution processes of quasi-isolated systems are viewed as sequences of states, where every state is considered as an initial state initiating the system's subsequent evolution, then it turns out to be possible to find order and law. What turns out is that, with a good choice of what is to be taken as a state for any quasi-isolated system, one can discover a law that, given *any* initial state, gives the state that evolves from it at *any* subsequent time. Such a law, since it is specifically concerned with evolution, is referred to as a *law of evolution*.

For an example let us return to the solar system. It turns out that the specification of the positions of all the planets and moons at any single time is insufficient for the prediction of their positions at later times. Thus the specification of states solely in terms of position is not a good one for the purpose of finding lawful behavior. However, the description of states by both the positions and the velocities of the planets and moons at any single time does allow the prediction of the state evolving from any initial state at any subsequent time. The law of evolution in this case consists of Newton's three universal laws of motion and law of universal gravitation.

So the reduction needed to enable the discovery of order and law in the natural evolution of quasi-isolated systems is the conceptual splitting of the evolution process into initial state and evolution. The usefulness of such a separation depends on the independence of the two "parts," on whether for a given system the same law of evolution is applicable equally to any initial state and whether initial states can be set up with no regard for what will subsequently evolve from them. Stated in other words, the analysis of the evolution process of a quasi-isolated system into initial state and evolution will be a reduction, if, on the one hand, nature indeed allows us, at least in principle, complete freedom in setting up the initial state, i.e., if nature is not at all concerned with initial states, while, on the other hand, what evolves from an initial state, once it is set up, is entirely beyond our control.

This reduction of evolution processes into initial state and evolution has proved to be admirably successful for everyday-size quasi-isolated systems and has served science faithfully for ages. Its extension to the very small seems quite satisfactory, although when quantum theory becomes relevant, the character of an initial state becomes quite different from what we are familiar with in larger systems. Its extension to the large, where we cannot actually set up initial states, is also successful. But we run into trouble when we consider the Universe as a whole. One reason for this is that the concepts of order and law are scientifically meaningless for the evolution of the Universe as a whole [M49, M52]. Another reason is that it is not at all clear whether the concept of initial state is meaningful for the Universe; I do not think it is [M51].

The symmetry that is implied by reduction into initial state and evolution follows immediately from the independence of the two "parts," as described two paragraphs above. On the one hand, laws of evolution are an aspect of nature that is immune to possible changes, the changes being changes in initial states. On the other hand, initial states are an aspect of nature that is immune to possible changes, where the changes are hypothetical changes in laws of evolution, in the sense that initial states can be set up with no regard for what will subsequently evolve from them. This symmetry, together with that implied by the reduction into quasi-isolated system and surroundings, is intimately related to the symmetry that is predictability, which is discussed in Section 9.7.

PROBLEM

A person's behavior upon meeting a strange dog depends not only on the impression the dog makes, but also on the person's past history of relations with dogs. A person would react differently if he or she had been attacked by a strange dog in the past or if he or she had always had friendly relations with dogs. Consider that situation with regard to possible analysis in terms of initial state and evolution.

9.6. Reproducibility as Symmetry

Science rests firmly on the dual foundation of reproducibility and predictability, or perhaps better, on the triple foundation of reproducibility, predictability, and reduction. Science is concerned with the reproducible and predictable phenomena of nature, and any phenomenon that is either irreproducible or unpredictable or both, lies outside the domain of concern of science. Reproducibility is the replicability of experiments in the same lab and in other labs, which makes science a common human endeavor, rather than, say, a collection of private, incommensurate efforts. It makes science as much as possible an objective, or at least intersubjective, endeavor. (Intersubjectivity means that even if we are not sure about the objectivity of science, i.e., about its independence of any kind of observer, at least all *human* observers agree about what is going on.) We now show that reproducibility is a symmetry, and we also show that reproducibility implies analogy.

Let us express things in terms of experiments and their results. Reproducibility is then commonly defined by the statement that the same experiment always gives the same result. But what is the "same" experiment? Actually each experiment, and we are including here even each run of the same experimental apparatus, is a unique phenomenon. No two experiments are identical. They must differ at least in time (the experiment being repeated in the same laboratory) or in location (the experiment being duplicated in another laboratory), and might, and in fact almost always do, differ in other aspects as well, such as in spatial orientation (since the Earth revolves and rotates). So when we specify "same" experiment and "same" result, we actually mean equivalent in some sense rather than identical. We cannot even begin to think about reproducibility without permitting ourselves to overlook certain differences, where those differences involve time or location as well as various other aspects of experiments.

Consider the difference between two experiments as being expressed by the change that must be imposed on one experiment in order to make it into the other. Such a change might involve temporal displacement, if the experiments are performed at different times. It might (also) involve spatial displacement, if they are (also) performed at different locations. If the experimental setups have different directions in space, the change will involve rotation. If they are in different states of motion, a boost (a change of velocity) will be involved. We might bend the apparatus. We might replace a brass part with a plastic one. Or we might measure velocity rather than pressure. And so on.

But not all possible changes are changes we associate with reproducibility. Let us list those we do. We certainly want temporal displacement, to allow the experiment to be repeated in the same lab, and spatial displacement and rotation, to allow other labs to perform the experiment. The motion of the Earth requires spatial displacement and rotation even for experiments performed in the same lab as well as velocity boosts for those performed at different times or locations. Then, to allow the use of different sets of apparatus, we need replacement by other materials, other atoms, other elementary particles, etc. Due to unavoidably limited experimental precision we must also include small changes in the conditions. And we also need changes in quantum phases, over which we have no control in principle. Those are the changes associated with reproducibility that I can think of.

Let us denote the set of all changes we associate with reproducibility—and add any I might have overlooked—by *REPRO*. We now define reproducibility as follows: Consider an experiment and its result, consider the experiment obtained by changing the original one by any change belonging to *REPRO*, and consider the result obtained by changing the original result in the same way. If the changed result is what is actually obtained by performing the changed experiment, and if that relation holds for all changes belonging to *REPRO*, we have reproducibility.

As an example, imagine some experiment whose result is a purple flash emanating from some point in the apparatus some time interval after the switch is turned on. Now imagine repeating the experiment with the same apparatus, in the same direction and state of motion relative to the Earth, etc., but $8\frac{1}{2}$ hr later and at a location 2.2 km east of the original location. If that purple flash now appears $8\frac{1}{2}$ hr later than and 2.2 km east of its previous appearance, we have evidence that the experiment might be reproducible. (As we know in this business, whereas a single negative result disproves reproducibility, no number of positive results can prove it. A few positive results make us suspect reproducibility; many will convince us; additional positive results will confirm our belief.)

Symmetry is materializing here; reproducibility is indeed a symmetry. We see that in this way: Consider a reproducible experiment and its result. Change it and its result together by any change belonging to *REPRO*. The pair (changed experiment, changed result) is in general different from the pair (original experiment, original result), but there is an aspect of the pairs that is immune to the change. That aspect is the relation, call it physicality, actuality, reality, or whatever, that *the result is what is actually obtained by performing the experiment.* Said in other words, the symmetry that is reproducibility is that, for any reproducible experiment and its result, the experiment and result derived from them by any change belonging to *REPRO* are also an experiment and its actual result.

Reproducibility implies analogy (see Section 8.4). The analogy is that the changed experiment is to the changed result as the original experiment is to the original result for all changes belonging to *REPRO*, with the relation that the result in each case is what is actually obtained by performing the experiment.

PROBLEM

Discuss the connection between reproducibility and observer–observed reduction.

9.7. Predictability as Symmetry

The other ingredient of the foundation of science, along with reproducibility and reduction, is predictability. Predictability means that among the phenomena investigated, order can be found, from which laws can be formulated, predicting the results of new experiments. Then theories can be developed to explain the laws. We now show that predictability, too, is a symmetry and that predictability implies analogy.

Here again we express things in terms of experiments and their results. Predictability, then, is that it is possible to predict the results of new experiments. Of course that does not often come about through pure inspiration, but is usually attained by performing experiments, studying their results, finding order, and formulating laws.

So imagine we have an experimental setup and run a series of n experiments on it, with experimental inputs $exp_1, \ldots, exp_n$, respectively, and corresponding experimental results $res_1, \ldots, res_n$. We then study those data, apply experience, insight, and intuition, perhaps plot them in various ways, and discover order among them. Suppose we find that all the data obey a certain relation, denote it R, such that all the results are related to their respective inputs in the same way. Using function notation, we find that $res_k = R(exp_k)$, $k = 1, \ldots, n$. That relation is a candidate for a law $res = R(exp)$ predicting the result res for *any* experimental input exp. Imagine further that this is indeed the correct law. Then additional experiments will confirm it, and we will find that $res_k = R(exp_k)$ also for $k = n + 1, \ldots$, as predicted. Predictability is the existence of such relations for experiments and their results.

For an example of that, consider again a given sphere rolling down a fixed inclined plane, with the experimental procedure of releasing the sphere from rest, letting it roll for any time t, and noting the distance d the sphere rolls in that time. Here t and d are playing the roles of *exp* and *res*, respectively. Suppose we perform ten experiments, giving the data pairs $(t_1, d_1), \ldots, (t_{10}, d_{10})$. We study those data and plot them in various ways. The plot of distance d_k against square of time interval t_k^2 looks like all ten points tend to fall on a straight line. That suggests the relation that the distance traveled from rest is proportional to the square of the time interval, $d_k = bt_k^2$, $k = 1, \ldots, 10$. This suggests the law $d = bt^2$ predicting the distance d for *any* time interval t. As it happens, that hypothesis is correct, and all additional experiments confirm it. The relation $d_k = bt_k^2$ is found to hold also for $k = 11, 12, \ldots$, i.e., also for data pairs $(t_{11}, d_{11}), (t_{12}, d_{12}), \ldots$. Thus the relation of distance to time interval is a predictable aspect of the setup.

That predictability is a symmetry can be seen as follows: For a given predictable experimental setup consider all the experiment-result pairs (*exp*, *res*) that have been, will be, or could be obtained by performing the experiment. Change any one of these into any other simply by replacing it. The changed pair is different from the original one in general, but the pairs possess an aspect that is immune to the change. That aspect is that *exp and res obey the same relation for all pairs*, namely, the relation *res* = $R(exp)$. Put in different words, the symmetry is that for any experiment and its result, the experiment and its result obtained by changing the experimental input obey the same relation as the original experiment and result.

Just as for reproducibility, predictability implies analogy (see Section 8.4): For a predictable experimental setup any experiment is to its result as any other experiment is to its result. The same relation *res* = $R(exp)$ holds for the input *exp* and result *res* in each case.

It might be enlightening to consider a case study here. Let us study the archetypical case of Kepler and his three laws of planetary motion. While pondering the many and various astronomical phenomena known to him, Kepler found a certain order within the general confusion. He found a classification of the motions of the celestial bodies, whereby the motions of the then known planets were assigned to one class and the motions of all the other bodies were assigned to another. The significant characterization of the former class was that in a heliocentric reference frame:

(1) the orbit of each body is an ellipse with the Sun at one of its foci;
(2) the radius vector of each body from the Sun sweeps out equal areas in equal time intervals; and
(3) the squared ratio of the orbital periods of any two bodies equals the cubed ratio of their respective orbital major axes.

The latter class was characterized, of course, by the nonfulfillment of (1)–(3). Thus the planetary motions were transformed, through the order Kepler perceived, from a set of individual motions, each requiring its own explanation, to a class of motions requiring a common explanation.

As we saw in Section 8.4, classification implies and is implied by analogy. Thus the motions of the known planets were found to be analogous in that they all fulfilled (1)–(3), while the motions of all the other celestial bodies were analogous in that they did not fulfill (1)–(3). We also saw that analogy is symmetry, and we just saw that this symmetry in the present case is the symmetry that is predictability. Kepler's order = classification = analogy suggested that (1)–(3) might be laws of planetary motion for all solar planets, not just for the then known ones, and indeed that hypothesis has been continually confirmed as additional planets have been discovered. To put things in terms of experiments and their results, the discovery of additional planets might be thought of as additional experiments, whose results fit the laws derived from the order perceived in the motions of the previously known planets, which might be thought of as previous experimental results. Kepler's laws inspired Newton to develop an explanatory theory in the form of his three universal laws of motion and his universal law of gravitation.

PROBLEMS

1. Discuss the connection between predictability and reduction to quasi-isolated system and surroundings and between predictability and reduction to initial state and evolution.

2. Does predictability imply reproducibility? Does the latter imply the former? Do they both imply each other? Does neither imply the other?

9.8. Symmetry at the Foundation of Science

We are now in a position to recognize and appreciate how large symmetry looms in the foundation of science. First, reproducibility, a major ingredient of that foundation, is a symmetry. And reproducibility implies analogy, which also is a symmetry. (See Section 9.6.) Second, predictability, another major ingredient of the foundation of science, is a symmetry too. And it, too, implies analogy, which again is a symmetry. (See Section 9.7.) And third, reduction, which can usefully be considered a third major ingredient of the foundation of science, is a symmetry. (See Section 9.2.)

9.9. Summary of Chapter Nine

We started out in Section 9.1 by defining nature as the material universe with which we can, or can conceivably, interact. Science, then, is our attempt to understand the reproducible and predictable aspects of nature. In Section 9.2 we saw that science operates by the method of reduction, by "slicing" the Universe into "parts," and that reduction implies symmetry. Three ways reduction is applied in science are: observer and observed, which was considered in Section 9.3; quasi-isolated system and surroundings, considered in Section 9.4; and initial state and evolution, in Section 9.5. The symmetry implied by each way of reduction was pointed out. It is for quasi-isolated systems that order and law are found.

Science rests firmly on the triple foundation of reproducibility, predictability, and reduction. In Section 9.6 we saw in detail what reproducibility, the possibility of replicating experiments, is, how reproducibility is a symmetry, and how it implies analogy. In Section 9.7 we saw the same for predictability, the existence of order from which laws can be formulated, predicting the results of new experiments. And putting it all together in Section 9.8, we saw that, via reproducibility, predictability, reduction, and analogy, symmetry rightly claims major importance in the foundation of science.

CHAPTER TEN

More Symmetry in Science

The role of analogy in science is considered. Two additional symmetries in science are discussed: symmetry of evolution of quasi-isolated systems and symmetry of states of systems.

As in the preceding chapters the discussion is conceptual, and the physical significance of the concepts is emphasized throughout. There is no formalism, and no special mathematical preparation is needed.

If you are on the "concept" track, after finishing this chapter you should proceed to Chapter 2.

10.1. Analogy in Science

Since analogy is symmetry, I would be remiss if I did not take this opportunity to elaborate a bit on the role of analogy in science. It has been stated [M34]: "The value of ... analogies in stimulating research ... [is] self-evident ...," and "... it cannot be denied that analogy plays an important role in scientific creativity." I would put it much more strongly and state that analogy is *essential* in *any* science-related activity. My justification for this claim is the following.

We show in Section 9.6 and 9.7 that analogy lies at the foundation of science, specifically that both reproducibility and predictability imply analogy. For reproducibility, any experiment–result pairs can be changed by any of the set of changes associated with reproducibility, and the changed result is what is actually obtained by performing the changed experiment. Thus all experiment–result pairs related by reproducibility-associated changes are analogous in that they all obey the same relation, namely, that the result is what is actually obtained by performing the experiment. For predictability, all actual and potential experiment–result pairs for the same predictable experimental setup are analogous in that they all obey the same law, usually expressed as a mathematical relation, as in the example of the rolling sphere.

Zooming in on predictability and order, the analogy involved there is more familiar than might be thought. After all, order, in any sense of the term, involves classification, which implies analogy. Thus, however one prefers to look at the matter, analogy is essential for any science-related activity, be it science teaching, science research, or scientific creativity.

For discussions of analogy in science see [M32], for a very general discussion, and [M34], where analogy symmetry is given the name "logical symmetry" and its pertinence to classification in science and to the periodic table of the chemical elements in particular is discussed.

Examples are certainly warranted here, and we might start with Kepler's three laws of planetary motion, which served us so well above. Kepler concluded from the astronomical data available to him that the motions of all the observed planets obeyed the same laws. That made the planets analogous, introducing order into the astronomers' picture of the solar system, and enabled Newton to derive his three laws of mechanics and law of gravitation.

For another example, Mendeleev's periodic table of the chemical elements (Dmitri Ivanovich Mendeleev, Russian chemist, 1834–1907) exhibited order and analogies among the elements [M34]. Those analogies were a major driving force in the development of models of atomic structure and in the development of quantum theory to explain the structure.

For yet another example, the analogies discovered among the so-called elementary particles [M38, M44, M66] were and are essential to the development of theories of their behavior and structure.

PROBLEMS

1. Describe in detail the analogies determined by the periodic table of the chemical elements.

2. What are some of the analogies among the elementary particles?

10.2. Symmetry of Evolution

The symmetries of reproducibility, predictability, and reduction indeed are essential for and inherent to science. However, they do not exhaust the usefulness and importance of symmetry for science. Two additional important and useful manifestations of symmetry in science are the symmetry of evolution of quasi-isolated physical systems and the symmetry of states of physical systems. First we discuss the former.

Systems evolve. Systems evolve from initial states to final states, if considered discretely, or, considered continuously, the states of systems are functions of time in general. Science is concerned with the evolution of quasi-isolated systems, because their evolution is found to have reproducible and predictable features. Indeed, the choice of what is to be considered a state for a system is made in such a way as to maximize those features.

(Note the circularity here. A system is declared quasi-isolated if it exhibits reproducibility and predictability with suitable choice of state. In fact, it is by the lack of such reproducibility and predictability that new effects and interactions are discovered, such as the discovery of the neutrino, for example.)

The symmetry of evolution of quasi-isolated systems, if it exists, means that there is some certain change that, if applied to any possible evolution, would result in another possible evolution, and if hypothetically applied to any impossible evolution, would result in another impossible evolution. (For quantum phenomena it would result in an evolution having the same probability.) The aspect of evolutions that is immune to the change is then their possibility or impossibility (or probability). Such a symmetry is reflected in the laws and theories describing and explaining the evolution of quasi-isolated systems and as such is also called a "symmetry of the laws of nature" [M53, M49]. Symmetries of evolution of quasi-isolated systems, or symmetries of the laws of nature, are intimately associated with conservations (also called "conservation laws") [M53, M30], such as the conservation of energy, the conservation of momentum, and the conservation of electric charge, but we do not elaborate that point here.

In the present discussion, too, it is convenient to express things in terms of experiments (initial states of quasi-isolated systems) and their results (the final states evolving from them). So consider some change applied to an experiment and to its result. If the actual result of the changed experiment is the same as the changed result of the original experiment and if that is valid for all experiments, we will have a symmetry of evolution (also a symmetry of the laws of nature). In other words, we will have a symmetry of evolution under a change, if for any experiment and its result, the experiment and result derived from them by that change are also an experiment and its actual result. The aspect of experiment–result pairs that is immune to the change is, just as for reproducibility, that *the result is what is actually obtained by performing the experiment.*

The analogy here should be clear: all experiment–result pairs connected by a symmetry of evolution are analogous in that they all obey the same relation, namely, that the result is what is actually obtained by performing the experiment.

The physical significance of symmetry of evolution is that nature is indifferent to certain aspects of physical systems, that the evolution of physical systems is independent of certain of their aspects, or stated in other words, that certain aspects of physical systems are irrelevant to the latter's evolution. All changes affecting only those aspects of physical systems that are irrelevant in that sense are symmetries of evolution, and any change affecting a relevant aspect cannot be a symmetry of evolution. So two systems differing only in irrelevant aspects will evolve in exactly the same way save for their (irrelevant) difference, and thus their difference will be preserved throughout their evolution, resulting in final states that differ precisely and solely as did the respective initial states. The change bringing about the difference is thus a symmetry of evolution.

When a symmetry of evolution is reflected in laws and theories as a symmetry of the laws of nature, it is sometimes referred to as expressing an "impotency." The idea is that the laws and theories are "powerless" to grasp and take into consideration the irrelevant aspects involved in that symmetry, those aspects to which nature is indifferent. Thus indifference of nature is exhibited as impotency of laws and theories.

Now, nature does indeed possess such symmetries. As far as is known at present, the universal symmetries of evolution of quasi-isolated systems are symmetries under: spatial displacements, temporal displacements, rotations, velocity boosts up to the speed of light (or, taken together and formulated group-theoretically, symmetry under the Poincaré group of space–time transformations (Jules Henri Poincaré, French mathematician, 1854–1912), the symmetry required by the special theory of relativity), and changes of phase. (Note that many of those symmetries are involved in the symmetry that is reproducibility, so that the very existence of science already implies certain symmetries of evolution, but we refrain from elaborating that point here.) Expressed in another way, the just-mentioned universal symmetries mean that with regard to the evolution of quasi-isolated systems nature does not recognize, respectively, absolute position, absolute instant, absolute direction, absolute velocity, or absolute phase.

As an example, symmetry of evolution under spatial displacements, the indifference of nature to position, can be expressed as: any two physical systems that are simultaneously in identical states, except for one being here and the other being there, will evolve into final states that are also simultaneously identical, except for one being here and the other there, respectively, for all heres and theres. Thus nature does not recognize absolute position through the evolution of quasi-isolated systems. Laws and theories must accordingly be impotent with regard to position, and the only position variables that can be allowed to enter them are relative positions.

As another example of symmetry of evolution, but not universal symmetry, consider any macroscopic system, which can be described both in terms of its macrostates and in terms of its microstates, where every macrostate represents a class of corresponding microstates. The macroscopic evolution of such a system is indifferent to the actual microstate realizing its macrostate, or in other words, its microstate is an irrelevant aspect of its macrostate with regard to its macroevolution. Thus we have symmetry of macroevolution under the change of microstates corresponding to the same macrostate.

To be more specific, imagine a quantity of ideal pure gas in thermal contact with a heat bath. The gas's macrostates can be specified by the quantities pressure, volume, temperature, and amount of gas (for instance, the number of molecules or moles). The macroevolution of the gas is described by the ideal gas law $pV = nR\Theta$, where p denotes the pressure, V the volume, n the quantity of gas, Θ the absolute temperature, and R is a suitable constant. The gas's microstates can be specified by the position and velocity of every one of its molecules. Its microevolution is described by Newton's laws of motion and whatever intermolecular force law applies. The heat bath by definition maintains a constant temperature, and we do not need to get into its microstates.

Now, imagine starting with the gas in some macrostate and the heat bath at twice the gas's temperature and putting the two in contact, while keeping a constant quantity of gas at a constant volume. Then the system will very rapidly evolve to a macrostate in which the gas has twice the absolute temperature and twice the pressure than it did before the process and in which the bath temperature is unchanged. That evolution is independent of the particular microstate the gas and the bath are in at any time. Whatever the combined system's initial microstate, it will evolve to some microstate corresponding to its final macrostate.

It is also a possibility, and in fact it happens, that nature possesses symmetries of evolution that are not valid for all experiments, but only for subsets of them. Such inexact symmetries are described, according to the case, as "partial," "limited," or "approximate." For example, symmetry of evolution under the change of particle–antiparticle conjugation is valid only for all experiments that do not involve neutrinos. For another example, symmetry of evolution under the combined change of particle–antiparticle conjugation and spatial inversion seems to be valid for all experiments that do not involve neutral kaons, while including those involving neutrinos.

PROBLEM

Formulate carefully, in terms of experiments and their results, the meaning of symmetry of evolution under: change of orientation, change of velocity, change of time.

10.3. Symmetry of States

Another important and useful manifestation of symmetry in science is the symmetry of states of physical systems. This manifestation would seem reasonably straightforward: We have symmetry of states, or a state is symmetric, if it is possible to change a state in a way that leaves some aspect of it intact. That, however, is opening a Pandora's box of triviality and boredom, since very many physical systems are sufficiently complex that there are very many physically trivial and uninteresting changes that leave intact very many physically trivial and uninteresting aspects of their states.

Well, if it is physical significance and interest we want, and as scientists that is certainly what we should want, we had better let nature guide us. And nature has spoken: Let the immune aspects of states be their irrelevant aspects in the sense of the preceding section, those aspects to which nature is indifferent and with regard to which the laws and theories are impotent, whether totally, partially, approximately, or to some limited degree. The changes under which states are symmetric, then, are the changes involved in the symmetries of evolution, again either exact symmetries or partial, approximate, or limited symmetries. Under those changes *the changed state is indistinguishable (or approximately indistinguishable) by nature from the original.* The analogy here should be obvious.

That being the situation, then how, one might very well ask, do the changed and original states differ at all, and just what changes were actually performed? Indeed, there is no difference between the changed and original states and, indeed, the changes are invisible—within the context of the quasi-isolated system for whose evolution the immune aspects of states are irrelevant, so that the state is symmetric under the changes. That purely and simply follows from our definitions. The states are distinguishable and the changes detectable only with respect to some suitable external system to which the changes are not applied. Such a system is a frame of reference for the changes.

As an example, consider the action of any of the universal symmetry changes mentioned in the preceding section, say spatial displacement, on any state of a quasi-isolated system. Within the system the changed and original states do not differ in any way that nature can distinguish; their difference is irrelevant. Put in more familiar terms, no experiment carried out wholly within the system can detect any difference between the states. (However, they do differ relative to a fixed coordinate system external to the system under consideration.) Or in other words, absolute position is undetectable within a quasi-isolated system.

For another example, replace "spatial displacement" by "boost" (change of velocity) in the previous example. The bottom line then becomes: Absolute velocity is undetectable within a quasi-isolated system. And so on for the other universal symmetry changes: rotations and phase changes and the undetectability within quasi-isolated systems of absolute direction and absolute phase.

As an additional example, consider an equilaterally triangular homogeneous metal plate. The system possesses symmetry under rotations by 120° and 240° about its center within its plane with respect to appearance and macroscopic physical properties. That means that any triplet of states mutually related by those rotations are indistinguishable by means of external appearance and macroscopic physical properties. External appearance involves the evolution of light waves impinging on and absorbed and reflected by the surface of the plate, which means that such triplets of states absorb and reflect light in the same way, as far as our visual perception is concerned. That is the basic immunity of unchanged external appearance. As for macroscopic physical properties, the states of such a triplet are actually microstates corresponding to the same macrostate, and the macroevolution involved in macroscopic physical properties cannot distinguish among them. That is the basic immunity here.

And another example: With respect to macroevolution every macrostate of a quasi-isolated macroscopic system is symmetric under change of microstate realizing that macrostate. The difference between microstates corresponding to the same macrostate is irrelevant to the macroevolution of the system. For instance, every macrostate of a gas (specified, say, by pressure, volume, temperature, and quantity of gas) can be realized by a very large of number of microstates (characterized by the position and velocity of every molecule). As far as quasi-isolated macroevolution of the gas is concerned, it is immaterial which of all those microstates realizes a macrostate. If the gas evolves from some initial macrostate to some final macrostate, then whatever its initial microstate realizing its initial macrostate, its microevolution will take it into some final microstate realizing its final macrostate.

PROBLEM

Describe the symmetry of states of quasi-isolated systems under velocity changes ("boosts") and under rotations.

10.4. Summary of Chapter Ten

In Section 10.1 we saw that since both reproducibility and predictability imply analogy, analogy lies at the foundation of science.

Two additional manifestations of symmetry in science were discussed in Sections 10.2 and 10.3, respectively: symmetry of evolution of quasi-isolated physical systems, also called symmetry of the laws of nature, and symmetry of states of physical systems. The former involves an indifference of nature to some aspect of a situation, which is reflected in a corresponding "impotency" of scientific laws and theories. Symmetry of states is the indistinguishability by nature of different states. Both symmetries are related, since states of quasi-isolated systems that differ only in aspects to which nature is indifferent are indistinguishable by nature and evolve in essentially the same way.

Summary of Principles

Here is a summary of the six symmetry principles that are derived in this book. The name of the principle, the statement of the principle, a few words about it and its derivation, and the section in the book where it is derived are given for each one.

The Equivalence Principle

Roughly: *Equivalent causes—equivalent effects.*

Precisely: *Equivalent states of a cause → equivalent states of its effect.*

It is derived from the existence of causal relations in nature and from the character of scientific laws as expressions of those causal relations. The equivalence principle is fundamental to the application of symmetry in science. See Section 5.2.

The Symmetry Principle

Roughly: *The effect is at least as symmetric as the cause.*

Precisely: *The symmetry group of the cause is a subgroup of the symmetry group of the effect.*

It is derived directly from the equivalence principle and, of the two principles, is the one that is usually used. See Section 5.3.

The Equivalence Principle for Processes

Equivalent states, as initial states, must *evolve into equivalent states, as final states, while inequivalent states* may *evolve into equivalent states.*

It is derived by applying the equivalence principle to natural evolution processes of quasi-isolated physical systems. See Section 7.2.

The Symmetry Principle for Processes

The "initial" symmetry group (that of the cause) is a subgroup of the "final" symmetry group (that of the effect).

It is derived from the equivalence principle for processes and is essentially the application of the symmetry principle to natural evolution processes of quasi-isolated physical systems. See Section 7.2.

The General Symmetry Evolution Principle

For a quasi-isolated physical system the degree of symmetry cannot decrease as the system evolves, but either remains constant or increases.

It follows immediately from the symmetry principle for processes. The general symmetry evolution principle has theoretical significance, but is so general as to be quite useless. See Section 7.2.

The Special Symmetry Evolution Principle

Usually: *The degree of symmetry of the state of a quasi-isolated system cannot decrease during evolution, but either remains constant or increases.*

Equivalently: *As a quasi-isolated system evolves, the populations of the equivalence subspaces (equivalence classes) of the sequence of states through which it passes cannot decrease, but either remain constant or increase.*

This is the useful symmetry evolution principle. It is derived from the equivalence principle for processes with the additional assumption of nonconvergent evolution. See Section 7.3.

Onward

That brings to a close our introduction to the general theory of symmetry in science, and you should be well started on your way toward an understanding and appreciation of symmetry and its application in science. Keep in mind, though, that this was only an introduction; there is much more ahead. The fields where symmetry considerations have traditionally played an important role are quantum theory and crystallography. More recently symmetry considerations have grown in importance, sometimes considerably, in many additional fields, such as solid-state physics, nuclear physics, the physics of particles and fields (also called high-energy or elementary-particle physics), general relativity, physical chemistry, especially quantum chemistry, low-temperature physics, and thermodynamics. And even in less fundamental fields, such as biology and engineering, symmetry considerations are gaining in importance. I doubt whether there is a single field of scientific endeavor in which symmetry considerations cannot prove useful.

For more advanced study you are referred to the Bibliography, and many additional references can be found in the books and articles listed there. For additional not necessarily very more advanced reading see [S31], [S22], and [S28], the third of which includes a discussion of symmetry of composite systems.

Although our approach to symmetry in this book has been very "materialistic," in that we have studied the application of symmetry to science, i.e., to material systems and the laws governing them, you must not leave with the impression that this is all there is to symmetry. Symmetry has its "spiritual" sides too, especially symmetry in art. The wide applicability of symmetry is emphasized in [S22], and symmetry in art is discussed in [S31] and [S28]. See also [S26].

Bibliography

The principal purpose of this bibliography is to offer sources for parallel, supplementary, complementary, and subsequent reading. It consists of a compilation of books and articles that I have read, come across, read reviews of, seen references to, or written and consider suitable in content and level. No claim is made for completeness. Its makeup reflects may own view of the field. The entries are classified, in some cases rather arbitrarily, into four categories:

G. *Abstract Group Theory*. Little symmetry, if any; rare application; corresponds fairly well to Chapters 2 and 3.

S. *General Symmetry*. Usually little group theory, if any; no detailed application, such as that involving group representations; roughly corresponds to Chapters 4 and 8, with some relevance to Chapters 5, 9, and 10.

A. *Applied Symmetry and Group Theory*. Sometimes with a lot of abstract group theory; most of the entries are presentations of the type for which the present book is intended to supply the theoretical symmetry foundations; thus this category mostly offers sequels to Chapter 5, but some of the entries are good for Chapter 5 itself.

M. *Miscellaneous*. Not classifiable under G, S, or A; some relevance to Chapters 9 and 10; material relevant to Chapters 6 and 7 is mostly found here.

References to entries have the form [Xi], where X = G, S, A, M and $i = 1, 2, \ldots$, indicating the ith entry in category X.

For a broader and more extensive bibliography up to about 1980 see my article "Resource letter SP-2: Symmetry and group theory in physics" [S23]. In spite of its title it is not limited strictly to physics and covers also chemistry, biology, and art.

G. Abstract Group Theory

[G1]   P.S. Alexandroff, *An Introduction to the Theory of Groups* (Blackie, Glasgow, 1959).

[G2]   B. Baumslag and B. Chandler, *Theory and Problems of Group Theory* (Schaum's Outline Series, McGraw-Hill, New York, 1968).

[G3]   A.W. Bell, *Algebraic Structures—Some Aspects of Group Structure* (Allen & Unwin, London, 1966).

[G4]   G. Birkhoff and S. MacLane, *A Survey of Modern Algebra* (Macmillan, New York, 1965).

[G5]   F.J. Budden, *The Fascination of Groups* (Cambridge University Press, Cambridge, 1972).

[G6]   H.S.M. Coxeter and W.O.J. Moser, *Generators and Relations for Discrete Groups* (Springer-Verlag, Berlin, 1965).

[G7]   J.D. Dixon, *Problems in Group Theory* (Dover, New York, 1973).

[G8]   R. Gilmore, *Lie Groups, Lie Algebras, and Some of Their Applications* (Wiley, New York, 1974).

[G9]   I. Grossman and W. Magnus, *Groups and Their Graphs* (Random House, New York, 1964).

[G10]   C.A. Hollingsworth, *Vectors, Matrices, and Group Theory for Scientists and Engineers* (McGraw-Hill, New York, 1967).

[G11]   W. Ledermann, *Introduction to Group Theory* (Barnes & Noble, New York, 1973).

[G12]   E.S. Lyapin, A.Ya. Aizenshtat, and M.M. Lesokhin, *Exercises in Group Theory* (Plenum, New York, 1972).

[G13]   The Open University, *Groups I. Mathematics Foundation Course Unit 30* (Open University Press, Milton Keynes, UK, 1971).

[G14]   The Open University, *Groups II. Mathematics Foundation Course Unit 33* (Open University Press, Milton Keynes, UK, 1972).

[G15]   The Open University, *Groups Axioms, Group Morphisms. Topic in Pure Mathematics Units 4 and 5* (Open University Press, Milton Keynes, UK, 1973).

[G16]   The Open University, *Group Structure. Topics in Pure Mathematics Units 11 and 12* (Open University Press, Milton Keynes, UK, 1973).

[G17]   E.M. Patterson and D.E. Rutherford, *Elementary Abstract Algebra* (Oliver & Boyd, Edinburgh, 1965).

[G18]   A.A. Sagle and R.E. Walde, *Introduction to Lie Groups and Lie Algebras* (Academic Press, New York, 1973).

[G19]   G. Stephenson, *Matrices, Sets and Groups* (American Elsevier, New York, 1966).

[G20]   D.A.R. Wallace, *Groups* (Allen & Unwin, London, 1974).

[G21]   H. Weyl, *The Classical Groups, Their Invariants and Representations* (Princeton University Press, Princeton, 1946).

[G22]   T.A. Whitelaw, *An Introduction to Abstract Algebra* (Blackie, Glasgow, 1978).

S. General Symmetry

[S1]   I. Bernal, W.C. Hamilton, and J.S. Ricci, *Symmetry, A Stereoscopic Guide for Chemists* (Freeman, San Francisco, 1972).

[S2]   B. Bunch, *Reality's Mirror: Exploring the Mathematics of Symmetry* (Wiley, New York, 1989).

[S3]   R. Caillois, *La Dissymétrie* (Gallimard, Paris, 1973).

[S4]   H.S.M. Coxeter, *Regular Polytopes* (Dover, New York, 1963).

[S5]   H.S.M. Coxeter, *Introduction to Geometry* (Wiley, New York, 1969), Chaps. 2–5, 7, 10, 11, and 15.

[S6]   L. Fejes Tóth, *Regular Figures* (Pergamon, Oxford, 1964).

[S7]   V. Fritsch, *Left and Right in Science and Life* (Humanities, Atlantic Highlands, NJ, 1968).

[S8]   M. Gardner, *The New Ambidextrous Universe: Symmetry and Asymmetry from Mirror Reflections to Superstrings* (Freeman, New York, 1991).

[S9]   L. Glasser, "Teaching symmetry, the use of decorations," *J. Chem. Educ.* **44**, 502–511 (1967).

[S10]   B. Gruber et al. (editors), *Symmetries in Science I–VI* (Plenum, New York, 1980, 1986, 1989, 1990, 1991, 1993).

[S11]   I. Hargittai (editor), *Symmetry: Unifying Human Understanding* (Pergamon, Oxford, 1986); *Symmetry 2: Unifying Human Understanding* (Pergamon, Oxford, 1989).

[S12]   I. Hargittai and M. Hargittai, *Symmetry Through the Eyes of a Chemist* (VCH, Weinheim, 1986).

[S13]   A. Holden, *Shapes, Space, and Symmetry* (Columbia University Press, New York, 1971).

[S14]   H.R. Jacobs, *Geometry* (Freeman, San Francisco, 1974), Chap. 5.

[S15]   M.A. Jaswon, *An Introduction to Mathematical Crystallography* (American Elsevier, New York, 1965).

[S16]   G. Karl, "On geometrical symmetry," *Amer. J. Phys.* **35**, 98–101 (1967).

[S17]   H. Liebeck, *Algebra for Scientists and Engineers* (Wiley, New York, 1969).

[S18]   E.H. Lockwood and R.H. MacMillan, *Geometric Symmetry* (Cambridge University Press, Cambridge, 1978).

[S19]   A.L. Loeb, *Color and Symmetry* (Wiley, New York, 1971).

[S20]   J. Nicolle, *La Symétrie et Ses Applications* (Albin Michel, Paris, 1950); *La Symétrie* (Presses Universitaires de France, Paris, 1957).

[S21]   D. Park, "Resource letter SP-1 on symmetry in physics," *Amer. J. Phys.* **36**, 577–584 (1968).

[S22]   J. Rosen, *Symmetry Discovered: Concepts and Applications in Nature and Science* (Cambridge University Press, Cambridge, 1975).

[S23]   J. Rosen, "Resource letter SP-2: Symmetry and group theory in physics," *Amer. J. Phys.* **49**, 304–319 (1981); reprinted in [M50], pp. 1–16.

[S24]   J. Rosen, "Fundamental manifestations of symmetry in physics," *Found. Phys.* **20**, 283–307 (1990).

[S25]   J. Rosen, "Symmetry at the foundations of quantum theory," *Found. Phys.* **21**, 1297–1304 (1991).

[S26]   M. Senechal and G. Fleck (editors), *Patterns of Symmetry* (University of Massachusetts Press, Amherst, 1977).

[S27]   A.V. Shubnikov, N.V. Belov, et al., *Colored Symmetry* (Pergamon, Oxford, 1964).

[S28]   A.V. Shubnikov and V.A. Koptsik, *Symmetry in Science and Art* (Plenum, New York, 1974).

[S29]   A.F. Wells, *The Third Dimension in Chemistry* (Oxford University Press, Oxford, 1956).

[S30]   M.J. Wenninger, *Polyhedron Models for the Classroom* (National Council of Teachers of Mathematics (U.S.), Reston, VA, 1966); *Polyhedron Models* (Cambridge University Press, Cambridge, 1971).

[S31]   H. Weyl, *Symmetry* (Princeton University Press, Princeton, 1952).

[S32]   E.P. Wigner, "Violations of symmetry in physics," *Sci. Amer.* **213** (6), 28–36 (December 1965).

[S33]   I.M. Yaglom, *Geometric Transformations*, Vols. I, II, III (Random House, New York, 1962, 1968, 1973).

A. Applied Symmetry and Group Theory

[A1]   V. Bargmann, "Note on Wigner's theorem on symmetry operations," *J. Math. Phys.* **5**, 862–868 (1964); reprinted in [M50], pp. 147–153.

[A2]   S. Bhagavantam, *Crystal Symmetry and Physical Properties* (Academic Press, New York, 1966).

[A3]   S. Bhagavantam and T. Venkatarayudi, *Theory of Groups and Its Application to Physical Problems* (Academic Press, New York, 1969).

[A4]   G. Birkhoff, *Hydrodynamics, A Study in Logic, Fact, and Similitude* (Dover, New York, 1950).

[A5]   G.D. Birkhoff, "The principle of sufficient reason," *Rice Inst. Pamphlet* **28**, 24–52 (1941), reprinted in *Collected Mathematical Papers* (Dover, New York, 1968), Vol. 3, pp. 778–804.

[A6]   R.R. Birss, *Symmetry and Magnetism* (North-Holland, Amsterdam, 1964).

[A7]   D.M. Bishop, *Group Theory and Chemistry* (Oxford University Press, Oxford, 1973).

[A8]   A.D. Boardman, D.E. O'Connor, and P.A. Young, *Symmetry and Its Applications in Science* (Wiley, New York, 1973).

[A9]   H. Boerner, *Representations of Groups* (North-Holland, Amsterdam, 1963).

[A10]   H.A. Buchdahl, *An Introduction to Hamiltonian Optics* (Cambridge University Press, Cambridge, 1970), Chaps. 3–9.

[A11]   D.B. Chesnut, *Finite Groups and Quantum Theory* (Wiley, New York, 1974).

[A12]   C.D.H. Chisholm, *Group Theoretical Techniques in Quantum Chemistry* (Academic Press, New York, 1976).

[A13]   J.F. Cornwell, *Group Theory and Electronic Energy Bands in Solids* (North-Holland, Amsterdam, 1969).

[A14]   F.A. Cotton, *Chemical Applications of Group Theory* (Wiley, New York, 1963).

[A15]   A.P. Cracknell, *Applied Group Theory* (Pergamon, Oxford, 1968).

[A16]   A.P. Cracknell, *Group Theory in Solid State Physics* (Wiley, New York, 1975); also published as "Group theory in solid-state physics is not dead yet, *alias*, Some recent developments in the use of group theory in solid-state physics," *Adv. Phys.* **23**, 673–866 (1974).

[A17]   P. Curie, "Sur la symétrie dans les phénomènes physiques, symétrie d'un champ électrique et d'un champ magnétique," *J. Phys.* (3rd ser.) **3**, 393–415 (1894); reprinted in *Oeuvres de Pierre Curie* (Gauthier-Villars, Paris, 1908), pp. 118–141; English translation, J. Rosen and P. Copié, "On symmetry in physical phenomena, symmetry of an electric field and of a magnetic field," in [M50], pp. 17–25.

[A18]   G. Davidson, *Introductory Group Theory for Chemists* (Elsevier, London, 1971).

[A19]   J.D. Donaldson and S.D. Ross, *Symmetry and Stereochemistry* (International Textbook, London, 1972).

[A20]   P.B. Dorain, *Symmetry in Inorganic Chemistry* (Addison-Wesley, Reading, MA, 1965).

[A21]   J.P. Elliott and P.G. Dawber, *Symmetry in Physics* (Oxford University Press, New York, 1979).

[A22]   J.McL. Emmerson, *Symmetry Principles in Particle Physics* (Oxford University Press, Oxford, 1972).

[A23]   L.M. Falikov, *Group Theory and Its Physical Applications* (University of Chicago Press, Chicago, 1966).

[A24]   B. Felsager and C. Claussen, *Geometry, Particles and Fields* (Odense University Press, Odense, 1981).

[A25]   J.R. Ferraro and J.S. Ziomek, *Introductory Group Theory and Its Application to Molecular Structure* (Plenum, New York, 1969).

[A26]   G.G. Hall, *Applied Group Theory* (American Elsevier, New York, 1967).

[A27]   L.H. Hall, *Group Theory and Symmetry in Chemistry* (McGraw-Hill, New York, 1969).

[A28]   M. Hamermesh, *Group Theory and Its Application to Physical Problems* (Addison-Wesley, Reading, MA, 1962).

[A29]   D.C. Harris and M.D. Bertolucci, *Symmetry and Spectroscopy, An Introduction to Vibrational and Electronic Spectroscopy* (Oxford University Press, New York, 1978).

[A30]   V. Heine, *Group Theory in Quantum Mechanics* (Pergamon, Oxford, 1960).

[A31]   R.M. Hochstrasser, *Molecular Aspects of Symmetry* (Benjamin, New York, 1966).

[A32]   T. Inui, Y. Tanabe, and Y. Onodera, *Group Theory and Its Applications in Physics* (Springer-Verlag, New York, 1991).

[A33]   F.M. Jaeger, *Lectures on the Principle of Symmetry and Its Applications in All Natural Sciences* (Elsevier, Amsterdam, 1917).

[A34]   H.H. Jaffé and M. Orchin, *Symmetry in Chemistry* (Wiley, New York, 1965).

[A35]   L. Jansen and M. Boon, *Theory of Finite Groups. Applications in Physics: Symmetry Groups of Quantum Mechanical Systems* (North-Holland, Amsterdam, 1967).

[A36]   T. Janssen, *Crystallographic Groups* (North-Holland, Amsterdam, 1973).

[A37]   D.F. Johnston, "Group theory in solid state physics," *Rep. Progr. Phys.* **23**, 66–153 (1960).

[A38]   A.W. Joshi, *Elements of Group Theory for Physicists* (Halsted, New York, 1974).

[A39]   J. Kepler, *The Six-Cornered Snowflake* (Oxford University Press, Oxford, 1966, originally published 1611).

[A40]   M.F.C. Ladd, *Symmetry in Molecules and Crystals* (Wiley, New York, 1989).

[A41]   M. Lax, *Symmetry Principles in Solid State and Molecular Physics* (Wiley, New York, 1974).

[A42]   J.W. Leech and D.J. Newman, *How to Use Groups* (Halsted, New York, 1969).

[A43]   J.S. Lomont, *Applications of Finite Groups* (Academic Press, New York, 1959).

[A44]   G.Ya. Lyubarskii, *The Application of Group Theory in Physics* (Pergamon, Oxford, 1960).

[A45]   R. McWeeny, *Symmetry, An Introduction to Group Theory and Its Applications* (Pergamon, Oxford, 1963).

[A46]   P.H.E. Meijer and E. Bauer, *Group Theory: The Application to Quantum Mechanics* (North-Holland, Amsterdam, 1965).

[A47]   W. Miller, Jr., *Symmetry Groups and Their Applications* (Academic Press, New York, 1972).

[A48]   A. Nussbaum, *Applied Group Theory for Chemists, Physicists and Engineers* (Prentice-Hall, Englewood Cliffs, NJ, 1971).

[A49]   M. Orchin and H.H. Jaffé, *Symmetry, Orbitals, and Spectra (S.O.S.)* (Wiley, New York, 1971).

[A50]   L.F. Phillips, *Basic Quantum Chemistry* (Wiley, New York, 1965).

[A51]   J.J. Sakurai, *Modern Quantum Mechanics* (Benjamin/Cummings, Menlo Park, CA, 1985), Chaps. 4, 6, and 7.

[A52]   I.V. Schensted, *A Course on the Application of Group Theory to Quantum Mechanics* (NEO, Ann Arbor, MI, 1965).

[A53]   D.S. Schonland, *Molecular Chemistry, An Introduction to Group Theory and Its Uses in Chemistry* (Van Nostrand-Reinhold, New York, 1965).

[A54]   R. Shaw, "Symmetry, uniqueness, and the Coulomb law of force," *Amer. J. Phys.* **33**, 300–305 (1965); reprinted in [M50], pp. 27–32.

[A55]   M. Tinkham, *Group Theory and Quantum Mechanics* (McGraw-Hill, New York, 1964).

[A56]   W.-K. Tung, *Group Theory in Physics* (World Scientific, Singapore, 1985).

[A57]   A. Vincent, *Molecular Symmetry and Group Theory, A Programmed Introduction to Chemical Applications* (Wiley, New York, 1977).

[A58]   B.L. van der Waerden, *Group Theory and Quantum Mechanics* (Springer-Verlag, Berlin, 1974).

[A59]   H. Weyl, *The Theory of Groups and Quantum Mechanics* (Dover, New York, 1931).

[A60]   E.P. Wigner, *Group Theory and Its Application to the Quantum Mechanics of Atomic Spectra* (Academic Press, New York, 1959).

[A61]   A.B. Wolbarst, *Symmetry and Quantum Systems* (Van Nostrand-Reinhold, New York, 1977).

[A62]   B.G. Wybourne, *Symmetry Principles in Atomic Spectroscopy* (Wiley, New York, 1970).

[A63]   B.G. Wybourne, *Classical Groups for Physicists* (Wiley, New York, 1974).

M. Miscellaneous

[M1]   J. Aharoni, *Lectures on Mechanics for Students of Physics and Engineering* (Oxford University Press, Oxford, 1972), Chap. 23.

[M2]   H. Alfvén, *Worlds–Antiworlds: Antimatter in Cosmology* (Freeman, San Francisco, 1966); "Antimatter and cosmology," *Sci. Amer.* **216** (4), 106–114 (April 1967).

[M3]   J.M. Bailey, *Liberal Arts Physics: Invariance and Change* (Freeman, San Francisco, 1972), Section 2.

[M4]   H. Callen, "Thermodynamics as a science of symmetry," *Found. Phys.* **4**, 423–443 (1974).

[M5]   P.C.W. Davies, *The Physics of Time Asymmetry* (Surrey University Press, London, 1974).

[M6]   P.C.W. Davies, *Space and Time in the Modern Universe* (Cambridge University Press, Cambridge, 1977).

[M7]   P.C.W. Davies, *Other Worlds: A Portrait of Nature in Rebellion: Space, Superspace, and the Quantum Universe* (Simon and Schuster, New York, 1980).

[M8]   P.C.W. Davies, *The Forces of Nature* (Cambridge University Press, Cambridge, 1986).

[M9]   P.C.W. Davies and J.R. Brown (editors), *The Ghost in the Atom: A Discussion of the Mysteries of Quantum Physics* (Cambridge University Press, Cambridge, 1986).

[M10]   P.A.M. Dirac, "The evolution of the physicist's picture of nature," *Sci. Amer.* **208** (5), 45–53 (May 1963).

[M11]   M.G. Doncel, A. Hermann, L. Michel, and A. Pais (editors), *Symmetries in Physics (1690–1980)* (World Scientific, Singapore, 1988).

[M12]   G.H. Duffey, *Theoretical Physics: Classical and Modern Views* (Houghton Mifflin, Boston, 1973), Chapters 7–10.

[M13]   C.V. Durell, *Readable Relativity* (Harper & Row, New York, 1931).

[M14]   F.J. Dyson, "Mathematics in the physical sciences," *Sci. Amer.* **211** (3), 128–146 (September 1964).

[M15]   E.B. Edwards, *Pattern and Design with Dynamic Symmetry* (Dover, New York, 1967, originally published 1932).

[M16]   B. Ernst, *The Magic Mirror of M.C. Escher* (Tarquin, Stradbroke, 1985).

[M17]   M.C. Escher et al., *M.C. Escher, His Life and Complete Graphic Work* (Abrams, New York, 1982).

[M18]   G. Feinberg and M. Goldhaber, "The conservation laws of physics," *Sci. Amer.* **209** (4), 36–45 (October 1963).

[M19]   R.P. Feynman, *The Character of Physical Law* (MIT Press, Cambridge, 1965), esp. Chaps. 3–5.

[M20]   R.P. Feynman, R.B. Leighton, and M. Sands, *The Feynman Lectures on Physics* (Addison-Wesley, Reading, MA, 1963), Vol. 1, Chaps. 46 and 52.

[M21]   G. Gamow, "The exclusion principle," *Sci. Amer.* **201**(1), 74–86 (July 1959).

[M22]   G. Gamow, *One, Two, Three—Infinity: Facts and Speculations of Science* (Viking, New York, 1963).

[M23]   M. Gardner, "Can time go backward?" *Sci. Amer.* **216** (1), 98–108 (January 1967).

[M24]   M. Gardner, *The Relativity Explosion* (Random House, New York, 1976).

[M25]   M. Ghyka, *The Geometry of Art and Life* (Sheed & Ward, New York, 1946).

[M26]   M. Gourdin, *Lagrangian Formalism and Symmetry Laws* (Gordon & Breach, New York, 1969).

[M27]   W. Greiner and B. Müller, *Quantum Mechanics*, Vol. 2: *Symmetries* (Springer-Verlag, New York, 1991).

[M28]   H. Haken, *Synergetics—An Introduction: Nonequilibrium Phase Transitions and Self-Organization in Physics, Chemistry and Biology* (Springer-Verlag, Berlin, 1977).

[M29]   J. Hambidge, *The Elements of Dynamic Symmetry* (Dover, New York, 1967, originally published 1926).

[M30]   P. Havas, "The connection between conservation laws and invariance groups: Folklore, fiction and fact," *Acta Phys. Austriaca* **38**, 145–167 (1973); reprinted in [M50], pp. 108–130.

[M31]   R.A. Hegstrom and D.K. Kondepudi, "The handedness of the Universe," *Sci. Amer.* **262** (1), 108–115 (January 1990).

[M32]   M.B. Hesse, *Models and Analogies in Science* (University of Notre Dame Press, Notre Dame, IN, 1966).

[M33]   R.M.F. Houtappel, H. Van Dam, and E.P. Wigner, "The conceptual basis and use of the geometric invariance principles," *Rev. Mod. Phys.* **37**, 595–632 (1965); reprinted in [M50], pp. 54–91.

[M34]   W.B. Jensen, "Classification, symmetry and the periodic table," in [S11], Vol. 1, pp. 487–510.

[M35]   G. Kepes (editor), *Module, Proportion, Symmetry, Rhythm* (Braziller, New York, 1966).

[M36]   L.D. Landau and G.W. Rumer, *What Is Relativity?* (Fawcett, New York, 1972).

[M37]   T.D. Lee, "Space inversion, time reversal and particle–antiparticle conjugation," *Phys. Today* **19** (3), 23–31 (March 1966).

[M38]   T.D. Lee, *Symmetries, Asymmetries, and the World of Particles* (University of Washington Press, Seattle, 1988).

[M39]   J. Leite Lopes, *Lectures on Symmetries* (Gordon & Breach, New York, 1969).

[M40]   C.H. MacGillavry, *Fantasy and Symmetry: The Periodic Drawings of M.C. Escher* (Abrams, New York, 1976).

[M41]   R.A. Mann, *The Classical Dynamics of Particles: Galilean and Lorentz Relativity* (Academic Press, New York, 1974), Chaps. 3 and 6.

[M42]   E. Merzbacher, *Quantum Mechanics* (Wiley, New York, 1970), Chaps. 16 and 20.

[M43]   A. Messiah, *Quantum Mechanics* (North-Holland, Amsterdam, 1962), Vol. 2, Part 3.

[M44]   Y. Ne'eman and Y. Kirsh, *The Particle Hunters* (Cambridge University Press, Cambridge, 1986).

[M45]   J. Neuberger, "Rotation matrix for an arbitrary axis." *Amer. J. Phys.* **45**, 492–493 (1977).

[M46]   A. Palazzolo, "Formalism for the rotation matrix of rotations about an arbitrary axis," *Amer. J. Phys.* **44**, 63–67 (1976).

[M47]   T. Poston and I. Stewart, *Catastrophe Theory and Its Applications* (Pitman, London, 1978).

[M48]   P. Renaud, "Sur une généralisation du principe de symétrie de Curie," *C. R. Acad. Sci. Paris* **200**, 531–534 (1935); English translation, J. Rosen and P. Copié, "On a generalization of Curie's symmetry principle," in [M50], p. 26.

[M49]   J. Rosen, "Extended Mach principle," *Amer. J. Phys.* **49**, 258–264 (1981).

[M50]   J. Rosen (editor), *Symmetries in Physics: Selected Reprints* (American Association of Physics Teachers, Stony Book, NY, 1982).

[M51]   J. Rosen, "When did the Universe begin?" *Amer. J. Phys.* **55**, 498–499 (1987).

[M52]   J. Rosen, *The Capricious Cosmos: Universe Beyond Law* (Macmillan, New York, 1991).

[M53]   J. Rosen and Y. Freundlich, "Symmetry and Conservation," *Amer. J. Phys.*

46, 1030–1041 (1978); reprinted in [M50], pp. 135–146; and sequel, J. Rosen, "Symmetry and conservation: Inverse Noether's theorem and general formalism," *J. Phys. A* **13**, 803–813 (1980).

[M54]   E. Samuel, *Order: In Life* (Prentice-Hall, Englewood Cliffs, NJ, 1972).

[M55]   D. Schattschneider, *Visions of Symmetry: Notebooks, Periodic Drawings, and Related Work of M.C. Escher* (Freeman, New York, 1990).

[M56]   A. Sellerio, "Entropia, probabilità, simmetria." *Nuovo Cimento* **6**, 236–242 (1929); "Le simmetrie nella fisica," *Scientia* **58**, 69–80 (1935).

[M57]   J.N. Shive and R.L. Weber, *Similarities in Physics* (Wiley, New York, 1982).

[M58]   H. Stephani and M.A.H. MacCullum, *Differential Equations: Their Solution Using Symmetries* (Cambridge University Press, New York, 1990).

[M59]   P.S. Stevens, *Patterns in Nature* (Little, Brown, Boston, 1974).

[M60]   E.F. Taylor and J.A. Wheeler, *Spacetime Physics* (Freeman, San Francisco, 1966).

[M61]   R. Thom, *Structural Stability and Morphogenesis: An Outline of a General Theory of Models* (Benjamin, Reading, MA, 1975).

[M62]   D.W. Thompson, *On Growth and Form* (Cambridge University Press, Cambridge, 1942).

[M63]   B.C. van Fraassen, *Laws and Symmetry* (Oxford University Press, Oxford, 1989).

[M64]   L.L. Whyte, "Tendency towards symmetry in fundamental physical structures," *Nature* **163**, 762–763 (1949).

[M65]   E.P. Wigner, *Symmetries and Reflections* (MIT Press, Cambridge, 1967).

[M66]   C.N. Yang, *Elementary Particles* (Princeton University Press, Princeton, 1962).

[M67]   E.C. Zeeman, "Catastrophe theory," *Sci. Amer.* **234** (4), 65–83 (April 1976).

Index